Citizens and Rulers of the World

MAHSHID MAYAR

Citizens and Rulers of the World
The American Child and the Cartographic Pedagogies of Empire

The University of North Carolina Press *Chapel Hill*

This book was published with the assistance of the Authors Fund of the
University of North Carolina Press.

© 2022 The University of North Carolina Press
All rights reserved
Set in Arno Pro by Westchester Publishing Services
Manufactured in the United States of America

The University of North Carolina Press has been a member of the
Green Press Initiative since 2003.

Complete Library of Congress Cataloging-in-Publication Data is available at
https://lccn.loc.gov/2021052602.

ISBN 978-1-4696-6727-0 (cloth: alk. paper)
ISBN 978-1-4696-6728-7 (pbk.: alk. paper)
ISBN 978-1-4696-6729-4 (ebook)

Cover illustration: Boy on deck, *Harper's Young People 1885* (New York: Harper and Brothers, 1885), title page. From the author's collection.

A shorter, earlier version of Chapter 2 previously appeared as "From Tools to Toys: American Dissected Maps and Geographic Knowledge at the Turn of the Twentieth Century," in *Knowledge Landscapes North America,* ed. Christian Kloeckner, Simone Knewitz, and Sabine Sielke, American Studies: A Monograph Series, vol. 273 (Heidelberg: Universitätsverlag Winter, 2016), 99–118.

Citizens and Rulers of the World

MAHSHID MAYAR

Citizens and Rulers of the World
The American Child and the Cartographic Pedagogies of Empire

The University of North Carolina Press *Chapel Hill*

This book was published with the assistance of the Authors Fund of the University of North Carolina Press.

© 2022 The University of North Carolina Press
All rights reserved
Set in Arno Pro by Westchester Publishing Services
Manufactured in the United States of America

The University of North Carolina Press has been a member of the Green Press Initiative since 2003.

Complete Library of Congress Cataloging-in-Publication Data is available at https://lccn.loc.gov/2021052602.

ISBN 978-1-4696-6727-0 (cloth: alk. paper)
ISBN 978-1-4696-6728-7 (pbk.: alk. paper)
ISBN 978-1-4696-6729-4 (ebook)

Cover illustration: Boy on deck, *Harper's Young People 1885* (New York: Harper and Brothers, 1885), title page. From the author's collection.

A shorter, earlier version of Chapter 2 previously appeared as "From Tools to Toys: American Dissected Maps and Geographic Knowledge at the Turn of the Twentieth Century," in *Knowledge Landscapes North America*, ed. Christian Kloeckner, Simone Knewitz, and Sabine Sielke, American Studies: A Monograph Series, vol. 273 (Heidelberg: Universitätsverlag Winter, 2016), 99–118.

for [m], my favorite share of the alphabet
and to my students

Contents

Acknowledgments xi

Introduction: I Know by the Color 1

CHAPTER ONE
Growing Up and Going Far: Geography Primers, "Home Geography," and the World 25

CHAPTER TWO
Quiet as Mice: Dissected Maps, Domestic Fun, and the World in Pieces 61

CHAPTER THREE
A for Amoy, Z for Zanesville: Child-Made Geographical Puzzles, Finger-Tip Travelers, and Cartographic Intimacies of the World 90

CHAPTER FOUR
We Sing a Geography Song: The Writing Child, the Portable Home Front, and World Geography 129

Conclusion: Huckleberry Finn in the World 157

Notes 171
Bibliography 205
Index 229

Figures

1.1   David Harvey, "Grid of Spatial Practices" 19
1.1   Samuel Goodrich, "Going to tell about Geography" 31
1.2   Samuel Goodrich, "A Chinese selling Rats and Puppies for pies" 33
1.3   Samuel Goodrich, "Norwegian" 33
1.4   William Channing Woodbridge, "Comparative size and appearance of the Planets" 35
1.5   Samuel Goodrich, "Picture of the World" 36
1.6   Samuel Goodrich, "Picture of the World" 38
1.7   Sanford Niles, "Chart Showing the Distribution of the Races of Men" 39
1.8   H. Justin Roddy, "Map of the World Showing Colonial Possessions" 42
1.9   Edwin Grant Dexter, "Programme for Fully Graded System" 45
1.10 "Geography is the description of the surface of the earth, and its countries and their inhabitants" 53
1.11 Mary Howe Smith Pratt, "Races of Men" 54
1.12 Charles F. King, "Map of Pond" 58
2.1   John Spilsbury, *Europe Divided into Its Kingdoms* 64
2.2   Inside cover of the box of the *Series of Dissected Maps* 68
2.3   *Seat of War in the Island of Luzon* 72
2.4   Original box of *Clemens' Silent Teacher* 75
2.5   Box Cover of *Clemens' Silent Teacher* 77
2.6   *A New Dissected Map of the World with a Picture Puzzle of the Capitol at Washington* 82
2.7   *Up the Heights of San Juan: Our Boys Storming the Blockhouse in Front of Santiago* 83
2.8   *The Nations at Peace* 86

3.1	Sam Loyd, *Get Off the Earth Puzzle Mystery*	95
3.2 (a)	(left) Sam Loyd, *The Lost "Jap"*	96
3.2 (b)	(right) Sam Loyd, *Puzzle of Teddy and the Lion*	97
3.3	"Geographical Guessthestory," *Harper's Young People*	100
3.4	"River Puzzle," *Harper's Young People*	103
3.5	"Geographical Puzzle," *Harper's Young People*	105
3.6	"Alphabetical Cities," *Harper's Young People*	114
3.7	"Geographical Puzzle," *Harper's Young People*	115
3.8	Answer to "Geographical Puzzle," *Harper's Young People*	117
3.9	"Geographical Puzzle," *Harper's Young People*	120
3.10	"Pied Cities," *Harper's Young People*	123
3.11	"Hidden Rivers," *Harper's Young People*	124
4.1	List of children's names whose letters were not printed in *St. Nicholas*	130
4.2	Letter by Elizabeth Haviland Brown, *St. Nicholas*	153
C.1	Haroun Al Huck-El-Berri and the Seven Sages of Bagdad, *St. Nicholas*	160

Acknowledgments

As a genre, acknowledgments go beyond mere form. People and their generosity do not need to be formatted and referenced in conformity with this or that manual of style to be valid. Thus, free from the burden of endnotes or citations, my foremost thanks go to Angelika Epple and Heike Paul, who supported me in my encounters with the exciting challenges of completing a project that, from the outset, stood at the intersection of geographies, disciplines, and agencies. It was with their critical support and encouragement that I went through drafts of my chapters, changed my manuscript's title, and won against the chaos of endnotes. Even earlier than that, it was Robert W. Cherny who saw me as a historian in the first place. I can never thank him enough. Amy Kaplan opened the door to critical thinking in every single one of her works that I have encountered—this book is dedicated to her memory. Sabine Schaefer, Rita Gaye, Thomas Abel, and Thomas Welskopp—thanks for what the Bielefeld Graduate School in History and Sociology (BGHS) has been as a workspace because of your presence. And, at UNC Press, my sincere thanks to Mark Simpson-Vos, Catherine Hodorowicz, Dominique Moore, María Isela García, and the anonymous reviewers whose careful reading of my work helped it become a better version of itself.

Since my initial encounter with historical childhood in the Internet Archive, I have frequented a large number of archives and libraries. Early on, Uwe Spiekermann of the German Historical Institute Washington, D.C., made it possible for me to access the digitized copies of *St. Nicholas*, and I am ever thankful for his and his colleagues' generosity. It was with the financial support of the BGHS that I made it to the collections at the Library of Congress, the Cotsen Children's Library at Princeton University, and the Newberry Library, where months of digitizing sources and browsing collections revealed more than I could ever have wished for about the politics of archiving children's material. Also, sincere thanks to Sybille Jagusch, head of the Children's Literature Center at the Library of Congress, for her stimulating company and her guiding hand through the library's immense collections. Numerous archival trips, in person and online, funded by the German Research Foundation, the Robert Bosch Foundation, the Bavarian American Academy, the Fritz Thyssen Foundation, the Bielefeld Young Researchers'

Fund, and the International Youth Library have made the completion of this project possible. Digitized sources available on the internet pages of Harvard University, the Newberry Library, the Cotsen Children's Library, the University of Indiana, the New-York Historical Society, Georgetown University, the University of Southern Mississippi, the Strong National Museum of Play, the American Antiquarian Society, Project Gutenberg, Amherst College, and Google Books, as well as friends with the right IP address, spared me numerous trips to the U.S. embassy to apply for one-entry visas.

I am deeply indebted to Amy Kaplan, Ann Laura Stoler, David Harvey, Fredric Jameson, Hsuan L. Hsu, James R. Akerman, Joyce Chaplin, Karen Sánchez-Eppler, Megan Norcia, Martin Brückner, Neil Smith, Patricia Crain, Robin Bernstein, and Susan Schulten, whose brilliant works on empire, geography and cartography, children and childhood, pedagogies of empire, literacy, race and coloniality, and the historical archive as they see it introduced me to a wealth of maps and chapbooks, itineraries and toponyms, publishing houses and fantasylands, child diaries and doodles. Also, thanks to the Bielefeld University library, especially Gabriele Pendorf and Sabine Rahmsdorf, for acquiring the numerous volumes that I suggested and for having made the library a welcoming place on days when I could not work in the isolation of my office. An earlier, shorter draft of chapter 2 originally appeared in *Knowledge Landscapes North America*. I thank Christian Kloeckner, Simone Knewitz, and Sabine Sielke, the volume's editors, and the Winter Verlag for permission to reprint portions of that chapter here.

I do not have enough words to thank my colleagues and friends at the Department of British and American Studies at Bielefeld University, the International Youth Library, the German Historical Institute Washington, D.C., the Bavarian American Academy, the BGHS, and the German Association for American Studies for the wondrous communities they have been. Alexander Martin, Ann Laura Stoler, Anne Friedrichs, Brian Rozema, Christen Mucher, Cleovi Mosuela, Gleb J. Albert, Helen Gibson, Jessica Pliley, Katherine Aid, Patricia Skorge, Stephen Morgan, Vivian Gramley, and Wilfried Raussert read various drafts of my chapters at a time when I needed their edifying touches the most. Andres Cardona, Barrett Watten, Bettina Brandt, Brian Rouleaue, Christina Meyer, Djelal Kadir, Hedwig Richter, Jörg Bergmann, Julia Roth, Kritika Agarwal, Ludmilla Jordanova, Marc Priewe, Marion Schulte, Meike Zwingenberger, Mischa Honeck, Michelle Tiedje, Nicole Waller, Petra Woersching, Ruben Quaas, Sabine Meyer, Shahzad Bashir, Ulla Kriebernegg, and Wai Chee Dimock cheered me on through stimulating seminars, talks, and conference panels, or in informal exchanges. Sincere thanks to Amy

Leonard, Georgetown University, for hosting me and facilitating my first-ever archival research at the Library of Congress. And a heartfelt thank-you to Karen Sánchez-Eppler for her interest in my work and her hospitality during the time I spent at Amherst College as a postdoctoral visiting fellow.

During the time I worked on this manuscript, I had numerous friends by my side, people whose generosity with their time, sense of humor, and intellect made the writing process a delight. My friends in and out of academia (Afrooz Rajoul, Amir Sadeghipour, Annette Rukwied, Carmen Dexl, Cedric Essi, Diana Fulger, Dorsa Ghaemi, Elena Furlanetto, Elena Matveeva, Hamedeh Saghafi, Hasti Khodabakhsh, Hediyeh Nasseri, Helen Gibson, Hoda Badr, Jana Kristin Hoffmann, Jens Temmen, Judith Rauscher, Julia Andres, Julia Faisst, Julia and Raphael Susewind, Jürgen Lange, Katharina Hoß, Katharina Pohl, Maryam Armaghan, Masoomeh Rezaei, Matin Rahmandoost, Melanie Dejnega, Monika Bokermann, Özlem Tan, Parisa Assar, Samae Bagheri, Sarah-Lena Essifi, Yaatsil Guevara González, and Zoltan Simon) and my officemates over the years I spent with this book (Alexandra Nitz, Arne Käthner, Astrid Haas, Eduardo Relly, Li Sun, Luisa Ellermeier, Matti Steinitz, and Sune Bechmann Pedersen) turned this project into a process of (re-)making friends. Also, thanks to the wonderful students in my childhood studies and new empire studies seminars, whose inquisitive passion propelled me forward with the revisions. Brigitte Tlatlik, Christine Schmuckert, and Silvia Toma—I can never thank you enough for your open hearts and warm smiles. Azadeh, Cleovi, Katharina, Lili, Michelle, Narges, Niko, Pat, Sandra, Somayyeh, Stephen, Tyll, and Yaatsil—countless thanks for your friendship, the long chats, the much-needed walks, and the laughter.

I am further indebted to a wide number of objects and spaces: the mug I inherited from Li Sun; the table lamp Sune Bechmann gave me; the table globe that Masoud gifted me to help me survive the brain's storm; The Coffee Store, my favorite Bielefeld café; Hassan's tea-stand at Bielefeld University; my personal archive in the form of seven envelopes on my office wall; and all the means of procrastination—including books, computer games, movies, other writing projects, and the almighty Internet—made it possible for me to reside in familiar spaces while navigating the *terra incognita* of children's worlds.

And Masoud, who has been sitting next to me, now and always, joking about his prime role in the completion of this book—thank you for holding, sharing, being "the home."

Citizens and Rulers of the World

Introduction
I Know by the Color

There was one thing that kept bothering me, and by and by I says:

"Tom, didn't we start east?"

"Yes."

"How fast have we been going?"

"Well, you heard what the professor said when he was raging round. Sometimes, he said, we was making fifty miles an hour, sometimes ninety, sometimes a hundred; said that with a gale to help he could make three hundred any time, and said if he wanted the gale, and wanted it blowing the right direction, he only had to go up higher or down lower to find it."

"Well, then, it's just as I reckoned. The professor lied."

"Why?"

"Because if we was going so fast we ought to be past Illinois, oughtn't we?"

"Certainly."

"Well, we ain't."

"What's the reason we ain't?"

"I know by the color. We're right over Illinois yet. And you can see for yourself that Indiana ain't in sight."

"I wonder what's the matter with you, Huck. You know by the COLOR?"

"Yes, of course I do."

"What's the color got to do with it?"

"It's got everything to do with it. Illinois is green, Indiana is pink. You show me any pink down here, if you can. No, sir; it's green."

"Indiana PINK? Why, what a lie!"

"It ain't no lie; I've seen it on the map, and it's pink."

You never see a person so aggravated and disgusted. He says:

"Well, if I was such a numbskull as you, Huck Finn, I would jump over. Seen it on the map! Huck Finn, did you reckon the States was the same color out-of-doors as they are on the map?"

"Tom Sawyer, what's a map for? Ain't it to learn you facts?"

"Of course."

"Well, then, how's it going to do that if it tells lies? That's what I want to know."

"Shucks, you muggins! It don't tell lies."

"It don't, don't it?"

—Mark Twain, *Tom Sawyer Abroad*, 1894

Forced to board a mad scientist's balloon for a trip around the world, Tom, Huck, and Jim heatedly debate what maps are for.[1] The satirical conversation documents the boys' lack of experience as world travelers and their apparent unsuitability for the trip as poorly literate consumers of the most basic of tools such a trip would require: the world map. If they cannot read maps—if, in other words, they cannot agree as to how maps relate to landscape—then, how are they going to navigate the world?[2] *Tom Sawyer Abroad*'s characters' diverse patterns of consuming, relating to, and remembering the "truths" that maps represent turn cartography into a site of contention.[3] The boys' disagreement has its origins in the fact that, as Twain reminds us, they encounter and read cartographic maps differently and, as a result, hold dissimilar views about their surroundings and of where and how, in the interaction between maps and landscapes, geographical views are formed. Despite the entirely different conclusions that Twain makes them ultimately draw as to what a map displays, the stereotypical white American teenagers Huck and Tom—and, by proxy, the freed slave Jim—do concur on one point: in their indispensability as tools of modern life, maps "don't tell lies."[4] In effect, it is immediately after the boys naively agree that maps are incapable of lying that, in comparing maps with their not-yet-fully-formed perceptions of the external reality that maps stand for, their opposing readings of cartographic representations, scalar complexities, and coloring patterns clash.[5] Addressing his young, mainly white, readers, Twain establishes that maps can be just as polysemous and confusing as they are factual and enlightening.

In its brevity, the passage underscores the multiplicity of individuals' mental maps of seen and unseen landscapes. Unlike Tom, Huck insists that Indiana ought to be pink because his initial mental impression of that state was formed by a secondhand encounter with it through a colored paper map of the United States. Since Huck has put his trust in maps, for him pink (or blue or green) would precede the diverse spectrum of colors that the Indiana landscape offers when viewed from above. No doubt, Huck's totalizing, playful, almost fanatical trust in cartographic verisimilitude, his inability to understand the subtleties involved in the concept of cartographic representation, and his presumption of a one-to-one relationship between a map and the landscape it represents appear simplistic in contrast with the more nuanced geographical understanding that Tom demonstrates.[6] Tom, meanwhile, is incapable of further reasoning when confronted by Huck's question: "Tom Sawyer, what's a map for?" In the end, while still mostly disoriented, Tom and Huck agree that the purpose of a map is "to learn you facts."

Thanks to the rise of lithography earlier in the nineteenth century, maps had become so widespread that Huck's insistence that he had seen the color on "the map" confirms the ubiquity of maps in everyday American life, as a result of which, as Martin Brückner has observed, almost anyone could "cite [them] without the need to specify any particulars about the map's make, author, or publisher."[7] Perhaps a commentary on politics of space, Huck's belief that Indiana should be pink in reality because it is pink in its scaled-down representation on the map reminds us of the ways that, at the height of the age of empire, cartographic maps were as much about defining, even dictating, the relationship between spatial representation and external reality as they were about charting out spaces and tagging them with colonial toponyms and colors. As Jean Baudrillard writes of maps and mapping since the age of empire, "The territory no longer precedes the map, nor survives it. Henceforth, it would be the territory whose shreds are slowly rotting across the map."[8] Expanding on Baudrillard's observations, Thai historian Thongchai Winichakul (whose ideas informed Benedict Anderson's writing on colonial maps) has persuasively argued that once European empires rose to power, their maps began to precede, even suspend, external colonized spatial realities and to impose on them a new order as envisioned by the colonizer.[9] As Tom and Huck's debate confirms, by employing various scales and calculated patterns of absence and presence, cartographic maps masked, as much as they exposed, the nonlinear regimes of spatial knowledge production and the by-and-large disorienting relationships that groups of people were to develop with spaces and—through those spaces—with one another.

Right at the beginning of their involuntary world tour, Tom and Huck seem unable to relate to the map of their own country, let alone to that of the non-American world toward which they are bound. In Twain's portrait of the boys, they seem to have a hands-on, if playfully misleading, knowledge of cartography: they know it is fundamental to locating oneself, navigating domestic and foreign spaces, and determining the distance between one's place and other parts of the world. What is more, Twain echoes in the story American adults' ongoing concerns that children's geographic knowledge of the world is shifting, creative, and in urgent need of further formal training. More importantly—and I return to this point in chapter 1—as a politically minded American adult satirizing the limits of geographic knowledge among Americans (both children and adults), Twain points in this story to the still relatively peripheral state of world geography as a school subject, thus underscoring the significance of geographic literacy as a topic of national concern in the United States late in the century and questioning Americans' awareness of this problem.[10] On the whole, the story illustrates the significance of public understandings of geographic space and, specifically, of refining children's understanding of America's place in the world. As a satirical narrative, it further suggests that children's playful modes of adapting geographic knowledge produces unpredictable relationships to and conceptions of the world.

Spaces of Empire

From its debates over geographic literacy to its bleak and alarming portrayal of children as inexperienced consumers of cartography, *Tom Sawyer Abroad* sets the stage for this book, in which I examine the semantic career of geography and cartography as geotechnologies for teaching Americans about the world in the 1890s.[11] Reading Huck, Tom, and Jim as fantastic shadows, as fictionalized effigies of real American children in their ambivalent, intimate relationships with maps and instances of mapping, and exploring the wider political implications of the production and reception of world geography knowledge by Americans, *Citizens and Rulers of the World* interrogates the points of convergence between geopolitics, imperial literacy, and the imagined place of the young American nation in—and in relation to—the world at large.[12] Intrigued by the multilayered liaison between childhood and imperial pedagogy over the course of U.S. national history, I attempt in the following chapters to trace heatedly debated issues that Mark Twain—and, by extension, much of white America—instructed modern, youthful, and adventurous turn-of-the-century Americans to learn and become conversant about in their often perplexing encounters with a colossal, eclectic global space that had already been mapped, for the most part, by European colonizers.

In other words, *Citizens and Rulers of the World* labors to read the ways in which the U.S. empire found itself in the world against, but also aligned with, the ways American children looked around to find their way in maps: How and through what forms of geographic knowledge and cartographic material did American adults—professional geographers, writers of school geography books, mapmakers, fiction and travel writers, and toy manufacturers—experience, imagine, and narrate the world outside the continental borders of the United States? More importantly, how did they communicate the nation's transforming ideals, geopolitical aspirations, and cartographic urgencies to the next generation? In turn, to what extent was children's understanding of world geography formed and informed by what they learned at school, played with in the evening, and wrote about while reflecting on their trips around the globe (or on the world map)? Ultimately, how and by what means did children come to terms with the United States' changing geopolitical imperatives, draw playful sketches of the world, and narrate their own perceptions of the national, the imperial, and the global in spatial terms?

The turn-of-the-century United States faced numerous domestic issues. The closing of the western frontier, the backbreaking depression of the mid-1890s, the unprecedented increase in the number of foreign immigrants, concerns over socialist and communist tendencies, the challenge of sustaining an on-strike labor market, and the record mobility of the population—especially from rural to urban areas and from the South to the North—were among the forces that gave rise to pressing and complicated questions in American homes as well as in Congress, businesses, and schools.[13] Troubled by urgent domestic issues during the tumultuous final decade of the nineteenth century, Americans further reflected on their comparatively short history as a nation; their position as one nation among many in a larger world; the urgency to build more overseas commercial, military, and diplomatic outposts; and how to raise and educate children as subjects immersed in the ways of the empire. By this time, some Americans had over a century of nationhood—in the spirit of republicanism—and continental expansion—disguised as westward expansion—at their disposal as forces with which to "domesticate" newly added spaces and to integrate formerly non-American territories into the nation's geography through violently othering their earlier settlers.[14] Children of the 1890s grew up amid such nationwide spatial tension. Still coming to terms with the complexities arising at the *Treffpunkte* between the nation and its spatially unsettled empire, Americans emphasized geography as a key component of their national history, a mode of navigating time and space, and a means of making sense of themselves and the world. In fact, white Americans wrote their national history first and foremost through focusing on the nation's ever-changing geography and its national spatial be(com)ing. "What there was to tell of American progress," Myra Jehlen remarks, "was geographical, celebrated in the incarnation of the spirit of liberal idealism and the individualist self in the North American continent."[15] As the following chapters demonstrate, *Citizens and Rulers of the World* examines this geographical incarnation of the national spirit beyond the North American continent, drawing focus to the U.S. empire's unsettled global contours in the 1890s and beyond.

At the end of a century that put unprecedented emphasis on the human element in social sciences, politics, and natural sciences, and in light of the unique influence of industrialization and the ensuing urbanization and mobility of the human body across countries and between continents, geography also shifted, as a discipline, into the realms of the human and the political. Turn-of-the-century geography in the United States came, more than ever before, under heavy influence by European geography. This continental draft of geography had been developed by Alexander von Humboldt (1769–1859), Carl Ritter (1779–1859), and later Vidal de la Blache (1845–1918) and Friedrich Ratzel (1844–1904). As I discuss in more detail in chapter 1, in the final decades of the century, geography was undergoing sweeping changes, evolving from an amateur fad to an academic discipline that was taught at the most prestigious colleges and universities in Europe and the United States, and from an all-inclusive field of inquiry to a modern science with singular disciplinary boundaries, methodologies, and theories. At the same time, and toward the end of the nineteenth century, the nation's increasing emphasis on geography and spatial unity had roots in the rising concept of geopolitics. Simply put, geopolitics entailed an au courant reading of geography in political terms. Long at work in imperial dealings with the world, yet undertheorized by politicians and geographers of the time, geopolitics influenced even the most common definitions of geography. Geopolitik, in its traditional sense—that is, the influence of geography on the national and international policies of states—evolved out of the work of German zoologist, geographer, and ethnologist Friedrich Ratzel. Reconceived in this light, geography began in the 1880s and 1890s to be understood as the knowledge of spaces in relation to state power.[16] By this time, geography was no longer seen as "something already possessed by the earth but *an active writing of the earth*" (emphasis in original).[17]

The many shifts in the direction geography took were mostly visible in frequent deliberations about the trade-off between "human civilizations" and in the mostly taken-for-granted hierarchies that associated racial differences with landscape and climate in the works of German, French, British, and, later American geographers. American geographers and geologists such as Richard Elwood Dodge (1868–1952) and William Morris Davis (1850–1934), and even historians such as Frederick Jackson Turner (1861–1932), adopted variations of this post-Enlightenment, Eurocentric, relatively methodical understanding of geography as the study of the conditioning and competing interplay of the human, the social, and the natural, and ultimately of the ways various "civilizations" had evolved on regional and global scales.[18]

This shift in focus was further traceable, chapter 1 establishes, in the American schoolroom.[19] As the century waned, geography was taught at an unprecedented number of schools across urban and rural America as a stand-alone science, at times in relation with commerce or industry, and often free from religious and historical undertones. This American draft of geography as a discipline was heavily informed by vast conceptual shifts in the understanding of geography as well as by material changes in U.S. foreign policy, the nation's commercial priorities, and demands by professional geographers for resources to ensure the survival of nascent departments of geography at American colleges and universities. In response to these shifts and demands, professional organizations and educational committees, such as the Committee of Ten called for the revision and rewriting of geography schoolbooks. With professional geographers and geography teachers on board, and keeping an eye on the foundational changes introduced in secondary education in France, Britain, and Germany, the Committee of Ten convened in 1892 to assess the American high school curriculum, its weekly schedule, and the training of high school teachers as well as more general educational practices across the country. As the committee's exhaustive report, published in the spring of 1894, testifies, the committee took a strong stance on the responsibility of the United States as a modern nation to think about its educational system beyond the national level and to keep "the larger community of interests which now knit civilized peoples together" at its heart.[20]

Children of Empire

Here I wish to pause to briefly explore the pivotal place children and childhood occupy in this book's overall narrative, as my arguments revolve around a strong interdependence between childhood as an inevitable process of growing up and the rise of the United States to global empire as a tenacious, predominantly white project of going far. Both before and after the "invention" of modern childhood in the West (if indeed such a turning point can ever be determined with any degree of certainty), adults have held a wide range of views as to who a child is (and should be) and how childhood should be placed and mapped.[21] Since at least the start of the nineteenth century, and in the name of sheltering the cute and vulnerable figure of the child, adults have readily spatialized childhood and placed the child in the interiors of homes, orphanages, schoolrooms, and playgrounds, and, later in the twentieth century, in private bedrooms and amusement parks so much so that child vulnerability—and the necessity for the child to be (over)seen by adults—has synonymized childhood to domesticity. Following this, adults have either viewed childhood as a transitory biosocial stage of human life that is characterized by little or no power or attempted devising and coding childhood as a set of artifacts—in the form of education and entertainment—to be consumed by flesh-and-blood children. Indeed, ever since the Industrial Age capitalized on the commodification of concepts and objects alike, modern childhood has been understood by a great number of thinkers as an artifact in the making, a construct devised by adults: "Childhood is like a toy," write Joseph Hawes and N. Ray Hiner, "something designed by adults and given to children."[22] Put differently, generations of adults have coded and formed childhood as a versatile, multipurpose object that in turn is to be decoded, even de-formed, by children.[23] Adults have further implemented the construct of childhood as collective emotional leverage for social and political change or invoked it as a metaphorical reference to financial or political dependence (perceived to be symptomatic of poverty, insanity, enslavement, sexual "deviation," or colonizability).[24]

Already over a century old as a nation, the turn-of-the-century white Americans continued to discuss and reflect on definitions of childhood and children's unique modes of perception. On the one hand, brimming with ambivalence in approaching the nation as the ultimate materialization of an erstwhile romantic idea, a niche political argument, or an ongoing geopolitical project, Americans had come to believe that "the American was an old race through a young nation."[25] Before the end of the century, and overlooking the woes and wishes of the heterogeneous American public, privileged Americans—white and uniformly male—conceived of the nation in terms of youth without childhood, juvenility in want of political apprenticeship, and adulthood despite a comparatively short national history.[26] In 1904, the pioneering American youth psychologist Granville Stanley Hall (1844–1924) summed up these views in scientific language. To Hall, and to his readers among the more privileged portions of American citizenry, youth was a peculiar feature of white American national identity:

> In a very pregnant psychological sense ours is an unhistoric land. Our very Constitution had a Minerva birth, and was not the slow growth of the precedent.... Our literature, customs, fashions, institutions, and legislation were inherited or copied, and our religion was not a gradual indigenous growth, but both in spirit and forms were imported ready-made from Holland, Rome, England, and Palestine. To this extent we are a fiat nation, and in a very significant sense we have had neither childhood nor youth, but have lost touch with these stages of life because we lack a normal development history.... Our immigrants have often passed the best

years of youth or leave it behind when they reach our shores, and their memories of it are in other lands. No country is so precociously old for its years [as we are].[27]

In this sense, Americans seem to have believed that any reference to this whitewashed Native American–, Asian American–, and African American–free national history and its immediate future at the dawn of the new century was incomplete without reflecting on the juvenility of the nation, its dense commerce with the spatialized and commodified construct of childhood, and the role its young citizens were assigned as wardens of its future.[28]

At the same time, adult white male colonizers invoked the figure of the child in contexts other than the nation and in reference to the colonized—communities and individuals who were deemed incapable of sovereignty and self-governance. They called on childhood in attempts to justify colonization as a patriarchal, bifold system of burden and benevolence within the framework of "the family of nations," and deployed childhood as an excuse to distance themselves from allegedly racially inferior Others.[29] Indeed, over the past two to three decades, and predominantly informed by the precision tools of postcolonial studies, historians of childhood have examined the coproduction of nationhood and childhood and the deep historical semblances between parenting and colonization in various contexts. Perry Nodelman, for instance, has insightfully discussed the parallels between "Orientalist" worldviews as spelled out by Edward Said and the construction and cooptation of childhood in psychology and literature.[30] The inherent hierarchy at the heart of child-adult interactions lends itself particularly well to discourses that justify unequal adult-adult relations based on the logic of colonization, the politics of disenfranchisement, and so on.[31] While raising somewhat different questions than *Citizens and Rulers of the World*, what is at stake in this body of research is the commonly shared premise that adults understand childhood as a phase of plasticity and impressionability, and that this mode of understanding has led adult actors to interfuse the causes of childhood, nation or empire building, and territorial expansion/colonization.

Keeping the real flesh-and-blood child in view, adults have understood childhood as a phase of human life that needs to be outgrown in order for individuals to arrive at the "ideal" stage of adulthood, at which point they can (at last earn the potential to) be considered full-fledged historical actors. From this perspective, children are not generally viewed "as selves, but as stages in the process of making an adult identity—as if childhood could only be meaningful in retrospect."[32] Among the growing number of recent volumes that treat childhood as an artifact designed for real children to learn from, scholarship on primary and secondary educational reform in U.S. history is most noteworthy. The illuminating works by Patricia Crain, for one, examine the exposure of colonial and antebellum American children to literacy as part of projects that aimed at rearing ideal, patriotic adult citizens in the nation and the empire.[33] Offering a vivid examination of educational records such as hornbooks and ABC books, spellers, and works of fiction, Crain tells the story of American culture through the close attention she pays to the political potential of "the array of individual, social, and institutional practices surrounding the internalization of the alphabet, the first step in literacy training."[34]

Next to Crain's nuanced take on literacy both in conversation with American culture and in the realm of geographic knowledge production, a number of illuminating titles visit similar sites of material cultural production, including American school primers; among them, Martin Brückner's extensive research on the history of geography in early America, Neil Smith's insightful work on the globalization of the U.S. empire in cultural and geopolitical terms, and Susan Schulten's critical forays into the changing nature of American geography stand out.[35] Similarly, quite a few historians have turned their attention to the trade-off between childhood, entertainment, and national and imperial pedagogy.[36] Joining Martin Brückner's work on geography and didactics in early America, research by Ann McGrath, Megan Norcia, Rebecca Onion, and others establishes that empire and imperial role-play were constant elements in team sports, party games, and home theatricals, as if empire and rule-based children's play were limbs of the same body.[37] More recently, cutting-edge research by historians such as Norcia and Edlie Wong have focused specifically on geography games. In their work, they examine the games' representational relationship to imperial and national imaginations and didactics.[38] Norcia's article "Puzzling Empire" presents an important take on the relationship of dissected maps—and, more generally, of picture puzzles (invented in the eighteenth century)—to the geography and geopolitics of the British Empire. However, as chapter 2 demonstrates, given the number of geography games and puzzles produced and sold during the nineteenth century, and especially its closing decades, the field merits further critical attention in order to more fully understand their relationship to contested and competing versions of U.S. colonial and imperial pursuits, especially at the turn of the twentieth century.

As the discussions in this diverse and thought-provoking scholarly space indicate, children's history is all too often written by, for, and from the perspective of adults.[39] Even in historiography that draws direct lines between childhood and empire, children have often been dissociated from any degree of historical presence, while childhood has been implemented as a metaphorical tool to mark physical dependency, psychological deficiency, or political immaturity.[40] As Karen Sánchez-Eppler asserts in her examination of the multiple facets of historical American childhoods, "Inchoate, children are often presented as not yet fully human, so that the figure of the child demarcates the boundaries of personhood, a limiting case of power, voice, or enfranchisement. Hence for people who are not male, or white, or American, or considered sufficiently sane or sufficiently rich, exclusion from civil rights has often been implemented through analogies to the child."[41] Despite rapidly growing interest in understanding childhood as a historical phenomenon and in viewing children as historical actors or observers, to focus historical studies on children beyond research in such fields as the history of emotions, historical childhood studies, and feminist scholarship on mothering and motherhood was, until recently, frowned upon.[42] It is only in the past two to three decades that, thanks to extensive work in historical childhood studies, children have begun to be acknowledged as individuals who live in the present and have consequential encounters with great historical questions of their time, such as the geopolitical shifts of entire nations. The present volume joins this body of work and labors to interrogate some of the instances in which children's accounts of history and of growing up in an empire prove to be anything but flippant, derivative, or inconsequential.

Paying systematic archival attention to children as historical actors and observers, a great number of studies that record histories of childhood or deliver historiography based on children's actual historical experiences are uniquely political in tone and interdisciplinary in outlook. Be they focused on children's rights, definitions of childhood, questions of national identity or displacement, or children's views on and relationships to colonization, works by such historians as Steven Mintz, Phillip Hoose, and Ann Laura Stoler focus on real children and their experiences in various historical contexts.[43] This wave includes, additionally, works such as Jeff Bowersox's *Raising Germans in the Age of Empire*, Sánchez-Eppler's *Dependent States*, Mischa Honeck's *Our Frontier Is the World*, and the present volume, where children of the past are treated as worthy of extensive attention in relation to historical questions and events. *Raising Germans in the Age of Empire*, for one, studies the responses that real children fashioned toward the artifacts adults offered them to mold imperial childhoods.[44] Examining the rise of imperial Germany late in the nineteenth century, Bowersox maintains that much effort went into planning imperial agency among German children as they grew up alongside the emerging German empire. In the case of young empires such as the United States and Germany, the idea of empire and its accompanying motifs and images were carefully chosen and grafted onto juvenile material culture, entertainment, and pedagogy as matters of national consequence and imperial urgency. After all, as our works concur, modern imperialism has always been a longitudinal project that coincides with the growing up of two, if not more, generations. As Bowersox indicates in the case of Germany, and as I posit in the present volume in the case of the United States, to grow up with(in) a young imperial context was an experience that involved the input of several generations and many actors and that constituted a blend of adult expectations, geopolitical education, legitimizing role play, material and ideological consumerism, and appropriative engagement on the part of the nation's youth.

Indeed, recognizing that children are capable of meaningfully observing and interpreting historical events and, further, developing and examining archives of childhood are vital to a deeper, if always partial, understanding of the reasons why political systems such as empires—multigenerational power constellations devised, expanded, and guarded by adults—succeed or fail in sustaining their spirit across generations. Accordingly, *Citizens and Rulers of the World* approaches geographical material produced by children and adults as coextensive yet distinct and equally significant indices of American national identity around the turn of the twentieth century. Celebrating the fact that some minorities—women, working class, Black, indigenous, LGBTQIA+, etc.—have fought for and secured the right, at least in some communities and in certain historical contexts, to rewrite history from their own perspectives or to write their own share of local, national, and global histories, I aim at searching for childhood and children, the misplaced silhouettes in the historical archive, by raising questions that locate these presences, thus rendering these children and their historical afterlives more visible.[45]

Coordinates of Empire

With firm roots in historical childhood studies, *Citizens and Rulers of the World* further intervenes in recent discussions on the globalization of the U.S. empire, especially from a critical spatial point of view. As noted before, for the United States, the nineteenth century was one of well-guarded isolationism, "domestic" territorial expansion and commotion, heightened nationalist and nativist sentiments, and untiring efforts at posing as a united nation in the face of increased immigration and territorial acquisitions through negotiation and force. Such a state of history demanded a special focus on geography and on territorial unity.⁴⁶ After all, the nation's violent overland and overseas expansion westward, southward, and northward, and the policies with which the government justified seizing these territories from the Atlantic to the Pacific and from the Arctic to the equator, had turned the nineteenth century into a time of unprecedented spatial convulsion for the United States.⁴⁷ It was a century of national "coming of age," one which Hsuan L. Hsu characterizes as having been shaped through the "encroachment of vast, external spaces," which in turn initiated "vast geographical transformations."⁴⁸ At the same time, however, to simply assume that, by the end of the century, whoever knew geography knew the world is a misleading overstatement. After all, turn-of-the-century American children were born at a time when American adults were still busy grappling with, planning for, and "internalizing" those vast "external" spaces as their own.⁴⁹

As Edward Said has famously noted, one common, basic feature of empires—whether European, American, or Asian—during the age of empire was their territorial expansion onto already surveyed native land and other empires' colonies. However, while by the century's end older European empires, such as the British Empire, had clearly signposted their colonies as well as their regions of influence on maps and in interimperial political interactions, the United States was still looking for regions of influence outside its continental borders and beyond the Americas in general. The nation was negotiating the vocabulary, the power relations, and the arguments that it could build on in order to justify the ever-widening expanse of its empire beyond Manifest Destiny—what I refer to as the "spatially unsettled U.S. empire" and its spatial becoming.⁵⁰ "Archipelagic," a term introduced and theorized by Brian Russell Roberts and Michelle Ann Stephens, is a closely related term that captures the territorial nature of the ever-evolving, globalizing U.S. empire and its colonial expansionist policies beyond its continental borders, especially in the Pacific and the Caribbean.⁵¹ Resonant with my understanding of the U.S. empire as spatially unsettled, Roberts and Stephens define the archipelagic Americas as "the temporally shifting and spatially splayed set of islands, island chains, and island-ocean-continent relations which have exceeded US-Americanism and have been affiliated with and indeed constitutive of competing notions of the Americas since at least 1492."⁵²

In light of multicontoured spatial tensions and political commotion, President McKinley elaborated in his second inaugural speech (delivered on March 04, 1901) on how, at the end of a triumphant encounter with a European empire and its independence-seeking colonies in 1898, the world had changed to the benefit of the United States. Unprecedented in U.S. history and in contrast to the more domestic antiwar speech he had delivered in 1897, McKinley's 1901 speech posited Americans side by side with people from other parts of the world:[53] "The American people, intrenched [sic] in freedom at home, take their love for it with them wherever they go, and they reject as mistaken and unworthy the doctrine that we lose our own liberties by securing the enduring foundations of liberty to others."[54] Keeping silent on the Spanish-Cuban-Philippine-American War's ensuing controversies and ongoing conflicts, McKinley remarked: "We are now at peace with the world."[55] In reference to the country's overseas colonial possessions, he interpreted the course of American history as destined for global prominence, concluding that "after 125 years of achievement for mankind we will not now surrender our equality with other powers on matters fundamental and essential to nationality."[56] McKinley rejoiced in what he viewed as a newly-won equality with European empires, displaying resolution as the leader of the nation in keeping America's place and rank in "the family of nations."[57]

At the same time, and thanks to the rise of a widening class of "geographically privileged" Americans (academic geographers, professional cartographers, trained school geography teachers, and politicians)[58] an increasing percentage of Americans had begun, by the final quarter of the century, to make prolific use of maps for varying purposes, from visualizing the national census to keeping track of epidemics.[59] In November 1899, for instance, and celebrating the nation's entry into a century that *National Geographic* later called "the geographic century,"[60] President McKinley famously told an audience of missionaries from the Methodist Episcopal Church that, once his prayers to God about the "Filipino question" had been answered, the first presidential order that he had given was that "the chief engineer of the War Department (our mapmaker) ... put the Philippines on the map of the United States."[61] By this time, weaving politics ever more closely into the fabric of the everyday, maps had come to occupy an even more central place in American life than before, so much so that the Library of Congress opened a separate maps section—the Hall of Maps and Charts—in 1897, the same year in which the library opened its doors to the public.[62] Overall, as Brückner rightly asserts in his discussion of the changes in the world of cartography, maps that were produced in the United States "simultaneously hinged on local technologies and were globally intertwined."[63] Turn-of-the-century American children were born into an increasingly map-literate society in which cartography and the visual translation of undefined far-off places into charted-out, heterogeneous, renamed spaces—not only across the North American continent but across the world at large—had ceased to be the exclusive craft of geographers and the guiding tool of sailors.[64]

It was therefore inevitable that Americans experienced unprecedented levels of interest in the non-American world, an unparalleled urge to locate, historicize, and signpost their nation and their archipelagic "possessions" off their continental borders and in the world at large, and a mounting sense of curiosity and confusion in encountering it. One way for Americans to make sense of their first- and second-hand encounters with the world was, in Mark Twain's words, "a picnic on a gigantic scale."[65] In its original seventeenth-century sense—not remarkably different from its contemporary meaning except in its racial and social overtones—"picnic" implied collective, short-lived fun away from home and "civilization." It was considered a leisurely practice through which the bourgeoisie browsed, sampled, and appropriated allegedly less civilized spaces away from home—that is, the countryside, the distant landscape, the Holy Land, the Far East, and so on—in order to make new sense and use of those spaces in relation, and in contrast to, the "civilized" spatial order to which they felt entitled.[66] Reminiscent of the ways Americans had embraced the "exhibition culture" propelled by such international events as Chicago's 1893 World's Columbian Exposition, Twain's contemporaries were invited to view the task of exploring and mapping the world as a collective—that is, a national—pastime: the practice of traversing distances to other, "less civilized" climes, of passing through and consuming spaces and places.

In contrast to Twain's notion of "picnic" and outdoor fun, many Americans viewed their encounters with the world in much sterner terms. For a great number of Protestant Americans, the practice of encountering the non-American world was one of embarking on a Christian mission, an undertaking that went beyond the consumerist urges of the tourist age.[67] By the end of the nineteenth century, the United States had sent unparalleled numbers of Christian missionaries to all corners of the world. These encounters with a world that was spun into a web of colonization further indexed cultural influences that assumed the opposite direction— influences that can be understood as an aggressive "penetration ... of middle-class lived experience by this strange new global relativity of the colonial network."[68] Colonialism, as Fredric Jameson and others have indicated, introduced as much change into the lives of the citizens of the metropole as it tried to dictate—albeit in different ways—to the colonized. However, none of these perspectives, from Twain's picnic metaphor to Americans' urge to "[know] their place in global spaces" and the "civilizing" and Christianizing mission, can fully mask the true politics that the United States pursued.

Though they were latecomers to the global colonial network outside the Americas, Americans' interest in mapping the world and in further flagging their position in it stemmed, in part, from an urge to find new markets that would purchase the surplus of a national economy that was wrecked by successive waves of local and global economic depression and to identify new frontiers to settle and "civilize" in lieu of their about-to-close western frontier.[69]

Thus, in order to "know their place in global spaces" and to cope with the unprecedented disorientation that was the result of numerous encounters with the world at large, turn-of-the-century Americans found it crucial to "map [their] position in a mappable external world" by applying new spatial strategies and deciphering old spatial narratives.[70] The nation's full-scale initiation into highly entangled world relations by the century's end relied on cartography not only in its mathematical, scientific sense. Americans, both children and adults, were pushed to (re)map the world subjectively through real or imaginary travels and to locate themselves in it as Americans. Young turn-of-the-century America was at a point at which—according to Jameson—it had to acknowledge the inadequacy of earlier geographic wisdom and of the suitability of the Mercator projection and to supplement existing maps with increasingly complex, individualized cartographic technologies.[71] In other words—and this is one of this book's fundamental arguments—this was the age in which American politicians and educators, but also the wider American public, felt a need to appraise "figures of map and mapping... and try to imagine something else."[72]

With geographic literacy becoming nationally timely, if not individually imperative, American children and adults industriously created, read, modified, and redrew individual and collective "cognitive maps" of the world. As a practice, cognitive mapping, Christian Jacob sums up, "is a reading."[73] It is a reading arrived at by adopting spatial strategies that go beyond simply relying on cartographic maps to move between points A and B. Put differently, cognitive maps are mental maps that individuals and, by extension, communities develop over time in the negotiations between prior spatial knowledge (such as existing cartographic maps and atlases, traffic signs, address books, personal or political attachments to collective colonial pasts, demonstration or commemoration sites, and private addresses) and individual power-laden scenarios of movability and mobility in any given space. As a highly abstract and pliable internal process involved in maintaining such mental maps, Jameson's "cognitive mapping" points to the set of strategies that individuals implement in order to cope with navigating the unprecedented complexities of the spaces that they encounter, pass through, dominate, or inhabit.[74]

The term "mental mapping" was originally coined and defined by Kevin Lynch in regard to urban planning.[75] According to Lynch, individual subjects develop cognitive maps of where they live or spaces that they enter for the first time through signposting, landmarking, tagging significant symbols, and the like. In analyzing late capitalism, Jameson adopted Lynch's concept, redefined it in broader ideological terms, and added a deeper historical, analytical aspect to it in order to deal with complex social structures in national and global spaces. In Jameson's sense, cognitive mapping implies a pragmatic yet ideological, fluid yet fractured rereading of the cartographic map. It involves a reconfiguration and rescaling of the already scaled-down representations of space for individual use. However, Jameson's understanding of the concept fails to identify a systematic set of practical steps for realizing cognitive maps and deploying them when navigating spaces. In fact, as his critics agree, "Cognitive mapping is the least articulated but also the most crucial of the Jamesonian categories."[76]

What actual steps do individuals take as they draw or revise cognitive maps? What levels of interaction between selves, power systems, and spaces enable people to extrapolate from their immediate everyday experiences of space and place to form comprehensible and comprehensive cognitive maps of the world around them? How do individuals navigate spaces that—though allegedly overlapping and purportedly linearly expanding in a string of definable, nested scales—contract and expand in reaction to a plethora of known and unknown factors and on unpredictable, jaggedly related scales? If, according to Jameson, individuals can hardly make sense of the world due to inherent tensions between their immediate local experiences and the broader, vastly more complex, totalizing global structures enveloping them, how do human beings direct and justify their movements on the planet? What spatial strategies do they apply in drawing, reading, revising, and sharing maps of such global hyperspaces, and how do those strategies emerge and evolve in response to specific historical circumstances? These are some of the crucial questions that Jameson leaves unanswered.[77]

In response to these questions, I resort to David Harvey's multilayered frame of spatial practices as a host of possible scenarios in which individuals' not-always-successful quest to cognitively map their place in the world takes form. Agreeing with Jameson about the confusion of living in, depending on, passing through, and even leaving an imprint on global hyperspaces, but also trying to grapple with it by drafting the outline of a road map for survival, Harvey develops a "grid" of spatial practices wherein he offers a multilayered, flexible, practical yet by no means comprehensive chart in which he classifies human spatial practices for coping with and surviving in global hyperspaces (figure 1.1).[78] In combining the study of cognitive processes with that of concrete material practices, Harvey's grid proposes a matter-of-fact model that labors to identify an underlining pattern in the otherwise elusive fabric of cognitive maps. Informed by his reading of Michel De Certeau, Gaston Bachelard, and Michel Foucault and, most importantly, to his understanding of space as theorized by Henri Lefebvre, and reflexive of his Marxist approach to what he understands as the capitalist grip on time and space, his model drafts an outline of material practices that produce spaces (human experiences), practices that help us make sense of spaces (through the force of perception), and mental inventions and interventions (broadly understood as imagination) that bring new meanings and potentialities to individual and collective spatial practices.[79]

In this light, I examine the nature of the ever-evolving cognitive maps of the world arrived at via multiple socially formed yet individually performed practices of spatial "experience," "perception," and "imagination." Cognitive maps, I argue, enabled Americans, young and old, to locate themselves in the world's expanse as citizens of an aspiring world power and to navigate the world in order to produce, access, appropriate, make sense of, or dominate its various parts, developing and concretizing ideas about the commercial, ideological, and imperial role of the United States in relation to other empires, nations, and colonies.[80] Mindful of Harvey's own cautionary words about the grid's thoroughness, I work with his tripartite categories of spatial practice as a model open to both criticism and extension.[81] Spatial practices, including drawing and reading cartographic maps, are beyond doubt more closely bound to and conditioned by external sociopolitical changes and various actors' access to power than they are influenced by the extent to which local and global spaces actually change. In order to theorize and better place the questions posed earlier, I therefore expand on Harvey's grid by identifying additional examples of spatial practices, such as renaming places as constituents of colonial networks of power, tagging places on maps, dissecting maps into jigsaw puzzles in a tool-to-toy process, and switching between various representations of the planet in line with the changing geopolitical urges of the time. As I discuss more closely in the following chapters, these practices could be classed as instances of what Harvey understands as "spaces of representation": innovative measures that enabled American children and adults to imagine spaces outside "home" and then decide to either distance themselves from them or proximate themselves to them.

In applying Harvey's grid to analyzing the sources examined in the following chapters—that is, those sources designed for, used, or produced by Ameri-

	Accessibility and distanciation	Appropriation and use of space	Domination and control of space	Production of space
Material spatial practices (experience)	flows of goods, money, people labour power, information, etc.; transport and communications systems; market and urban hierarchies; agglomeration	land uses and built environments; social spaces and other 'turf' designations; social networks of communication and mutual aid	private property in land; state and administrative divisions of space; exclusive communities and neighbourhoods; exclusionary zoning and other forms of social control (policing and surveillance)	production of physical infrastructures (transport and communications; built environments; land clearance, etc.); territorial organization of social infrastructures (formal and informal)
Representations of space (perception)	social, psychological and physical measures of distance; map-making; theories of the 'friction of distance' (principle of least effort, social physics, range of a good, central place and other forms of location theory)	personal space; mental maps of occupied space; spatial hierarchies; symbolic representation of spaces; spatial 'discourses'	forbidden spaces; 'territorial imperatives'; community; regional culture; nationalism; geopolitics; hierarchies	new systems of mapping, visual representation, communication, etc.; new artistic and architectural 'discourses'; semiotics.
Spaces of representation (imagination)	attraction/repulsion; distance/desire; access/denial; transcendence 'medium is the message'.	familiarity; hearth and home; open places; places of popular spectacle (streets, squares, markets); iconography and graffiti; advertising	unfamiliarity; spaces of fear; property and possession; monumentality and constructed spaces of ritual; symbolic barriers and symbolic capital; construction of 'tradition'; spaces of repression	utopian plans; imaginary landscapes; science fiction ontologies and space; artists' sketches; mythologies of space and place; poetics of space; spaces of desire

Source: in part inspired by Lefebvre (1974)

FIGURE I.1 "Grid of Spatial Practices," David Harvey, *The Condition of Postmodernity: An Inquiry into the Origins of Cultural Change*, 1989, 220.

can children—I turn to global microhistory, a recent spin-off of Carlo Ginzburg's *Paradigma indiziario*, or the "evidential paradigm" proposed in his 1979 "Spie: Radici di un paradigma indiziario."[82] Ginzburg's approach to microhistory—a term promoted by subsequent generations of historians[83]—is one of detective work: systematic attention paid to "discarded clues" and overlooked details.[84] The evidential paradigm underscores the significance of locating, marking, examining, and thus reading easily ignorable or seemingly inconsequential clues.[85] Indeed, the "method of clues" allows historians not only to take seemingly irrelevant or mundane evidence seriously, but also to make credible claims about larger patterns and latent practices based on the micro-level analysis of those traces.[86] Within the frame of "relational global microhistory" developed by Angelika Epple, this method translates into a quest for tracing and interpreting discarded or overlooked "clues" on the individual or local level—including items made by, or events and circumstances surrounding, disempowered or marginalized actors such as children—in order to contribute to a more nuanced understanding of events, spatial arrangements, or processes that pertain to the collective, the national, and the global.[87] *Citizens and Rulers of the World* offers a close examination of such clues, actors, circumstances, spaces, and events through close textual and visual analysis of the primary sources for and by American children. In order to arrive at answers to questions about the emergence of U.S. overseas imperialism at the end of the nineteenth century, the analytical frame that I develop and employ in the following chapters entails practices that I term "a semiotics of the overlooked": meaning-making instances of cognitive mapping that read the contents of school primers, cartographic maps, and geographic toys designed for children (in chapters 1 and 2) alongside letters, geographical brainteasers, and puzzles composed by children (in chapters 3 and 4). As the following chapters illustrate, children expressed themselves as young Americans with the help of carefree but consequential spatial practices that mark the domestic, the immediate, and the personal next to, away from, within, and surrounding the foreign, the distant, and the global.

Archives of Empire

As a modern Western empirical science with roots in the Enlightenment, geography was initially put into practice through the colonial pursuits of European empires. In what follows, chapters 1 and 2 make the case that within almost a century, American geography had followed suit, redefining itself both as a science and as a set of self-positioning practices.[88] In the last one or two decades of the nineteenth century, geography—and especially world geography—began to be defined, modified, and even at times thematically revolutionized in professional and everyday spheres of American life, including primary and secondary education and popular culture. Taking up the violent and far from straightforward task of mapping the world and flagging their place on such maps, American adults continually urged the younger generations to learn about the non-American world. Geographic knowledge, American adults decided, was essential to youth—which in their view was going to secure the future of the empire. In this light, chapters 1 and 2 survey schoolbooks, map games, and geographical toys produced by geographically privileged adults in order to identify the forms of knowledge and competency that geographical educators sought to inculcate in future generations.

The sources explored in these chapters include material that primarily addressed well-to-do white American children—and children of color, orphans, and children of poorer backgrounds as hand-me-downs from their original, privileged owners—while they were indoors: reading, studying, doing homework, and playing at home and at school. Tracing the path that geography took from a patriotic school subject in the wake of the American Revolution to a "modern" science by the end of the nineteenth century, chapter 1 explores the many contours of late-nineteenth-century U.S. geography education around the concept of "home geography"—a pedagogical tool that focused on fostering local experiences and observations in order to convey a sense of the world as an eclectic collation of familiar spaces of comfort and civilization called "home." Stepping out of American schoolrooms, chapter 2 studies geography as child entertainment, specifically as it appeared in the form of dissected maps (an early form of jigsaw puzzles) and globe-trotting board games. Examining U.S.-manufactured dissected maps and picture puzzles in tandem with the imperatives of an ever more complex and globally involved United States, the chapter explores dissected maps' intended and unintended instructions, hints, and tips that tutored children in ways of seeing the larger world and imagining the United States as a central, or at least prominent, part of it. Taken together, books and toys lay bare the ways geography was deployed to teach American children how to appreciate, inherit, and run their expanding nation and its overseas empire. After all, with its power to invoke spatial imaginaries and to replace actual trips around the world for the young and the untraveled, geography was considered the "contemporary world-lore for the masses" and "the universology of the culturally poor."[89]

To further pursue this exercise, I turn in the second half of the book to the ways American children themselves engaged with diverse spatial practices of their own devising. Born into a world that had been surveyed and mapped by

European—and, to a lesser degree, American—colonizers, in which they encountered geography lessons during the day and were given geographical games and toys to play with in the evening, American children were encouraged to chart out the world for themselves. Busy with creative play, the quest for geographical signposting, and the search for unexplored lands, they had to imagine the full expanse of the already reconnoitered world before they could finally mark their homes in it. To make sense of these practices, I propose the term "cartography in progress"—a set of meaningful cognitive, visual, tactile, and emotional practices that, especially in the case of children playing with map games and dissected maps, outstripped the borders of factual geography, viewed the world beyond commercial or colonial terms, and troubled former geographical definitions of the world as a unitary space called planet Earth.

Taking a step closer to the slippery figure of the child, while also never fully departing from the world of adults, chapters 3 and 4 examine child-designed geographical puzzles as well as letters written by children detailing their trips to various parts of the world and their encounters with others. Delving into the subjective, elusive, and playful nature of children's geographical competencies, I argue in these chapters that while consuming what adults had designed for them, children engaged in a wide range of at times subversive spatial practices. Closely engaged in the semiotics of the overlooked, the chapters track the effects of seemingly unimportant events—such as a visit to a quarantine station off New York Harbor, a child's swimming experience in the Pacific, and a school fight in Weimar, Germany—and trace the responses of ostensibly marginal child actors to universal compulsory education in connecting the individual, the local, and the child-specific to the collective, the global, and the geopolitical.

Focusing on the geographical puzzles that children sent to the juvenile periodical *Harper's Young People* during the 1890s, chapter 3 focuses on what children "knew" about the world (in terms of random or alphabetically ordered lists of place names). While no evidence exists today of the original puzzles the periodicals received in order to establish their authorship, looking past these puzzles' literary aspirations and their entertaining and showy facades, I argue in this chapter that accounting for the scalar tension that characterizes children's playful relationship to the fundamentals of cartography as young puzzle makers is key to understanding how they produced geographic knowledge of a world that existed beyond and yet included the United States. Examining the letters that children sent to two of the most popular juvenile periodicals of the time, *St. Nicholas* and *Harper's Young People*, chapter 4 places emphasis on "what children perceived." In other words, I record in this chapter the many ways the writing child related to American and non-American worlds as practices in home geography.

In order to explore children's acts of cartography, I turn in these chapters to a more-or-less privileged demographic slice of American children. Consequently, chapters 3 and 4 investigate as their core case studies, the child-authored content of two juvenile periodicals that were at the peak of their popularity in the 1890s as their core case studies. Among hundreds of children's magazines launched in the United States,[90] about seventy-five English-language juvenile periodicals were in publication between 1885 and 1905.[91] Out of this teeming pool of postwar religious and liberal magazines aimed at children and young adults, I focus here on *St. Nicholas: Scribner's Illustrated Magazine for Girls and Boys* (1873–1943) and *Harper's Young People: An Illustrated Weekly* (1879–1899).[92] The offspring of the nation's two leading publishing houses, both magazines faced fierce competition from such established periodicals as the Boston-based *Youth's Companion* (1827–1929).[93] Yet historians agree that these two remarkably long-lived magazines were widely read in their time and have been highly revered in retrospect.[94] These periodicals targeted entire families. From the beginning, *St. Nicholas* had been envisioned as a companion to its adult counterpart, *Scribner's Monthly*: "They will be harmonious companions in the family."[95] Similarly, the House of Harper insisted that *Harper's Young People* be published in a "family of magazines," with multiple titles appropriate for readers of different ages.

Yet within the perimeters of the semiology of the overlooked that I pursue in this book, and while I avoid painting the child-adult relationship in an either-or light, in analyzing the largely unexplored archives of childhood, I distinguish between "publishing" and "printing" in a way that recognizes the disparate nature of making child- and adult-authored texts public; I use "publishing" when I examine the negotiations that routinely take place between professional writers and their editors and publishers, and I use "printing" to highlight the privileged yet far less powerful position of children as authors of their letters and puzzles in the adult-dominated periodical press. This distinction is meant to open up space for readerly/scholarly speculation. While adult writers could and did negotiate, accept, or reject certain changes in the texts they wrote, children's letters would be either thoroughly edited and modified for their content, spelling, and grammar without their consent and printed without their approval or printed with the original lapses, underscoring the newly literate children's perceived simplicity in thought and unstudied manner of composition.

DELVING INTO THE FAR-FROM-LINEAR, cross-generational exchanges and disruptions that characterize any and every political project as tenuous as empire, the present book is dedicated to the interrogation of the many ways turn-of-the-century American children—students, playmates and classmates, members of clearly marked socioeconomic classes and familial boundaries, readers of juvenile periodicals, and keen participants in nationwide epistolary networks—consumed geographic knowledge and produced spatial narratives and cognitive maps of their own. The close readings in the following chapters consider contexts wherein predominantly positively coded narratives about empire, expansionist policies, systemic racialization, national identity, patriotism, home, and so on were formed and re-formed, taught and learned, consumed pleasurably and passively, or challenged playfully and passionately. Eventually, while adults heralded the empire's global outreach by industriously circulating the idea and image of empire, scripting it in geography primers, magazines, toys, adventure literature, and cartography, it was children who inherited these narratives and acted as more than mere performers of the empire's scripts of fate and fiasco.

CHAPTER ONE

Growing Up and Going Far
Geography Primers, "Home Geography," and the World

In the wake of the American Revolution in 1776, and as the Constitution and the following Bill of Rights failed to mention public education, prominent New England figures such as Noah Webster, Abigail Adams, Benjamin Rush, and Jedidiah Morse expressed concerns about the republic's continuing ties with Europe. The new nation having ushered in an ostensibly didactic vacuum that had to be filled with knowledge produced *in* and *for* the United States, American educators, among others, emphasized the necessity of building up a national body of knowledge to be made accessible to the public. As Noah Webster remarked in 1790, citizens of the young republic needed to be educated from early childhood because "the Impressions received in early life, usually form the characters of individuals; a union of which forms the general character of a nation."[1] Apprehensive of the general character of the nation, Jedidiah Morse (1761–1826)—Congregationalist minister, Yale instructor, and "father of American geography"—warned about the schoolbook market, in which old and new volumes imported from England were still quite popular. During the first half-century following the War of Independence, American educators were particularly adamant about producing *American* textbooks. Morse called for the composition of geography books that focused on U.S. geography in order to correct the allegedly common errors in books composed in Europe. In the preface to *The American Geography*, believed to be the first "American" reader in geography, he blamed existing mistakes in the American geography primers of the time on European geography authors.[2] In turn, Morse promised to carry out his patriotic duty to educate not only Americans but also any non-Americans interested in American geography: "But since the United States have become an independent nation, and have risen into Empire," wrote Morse, "it would be reproachful for them to suffer this ignorance to continue; and the rest of the world have a right now to expect authentic information.[3] Indeed, if John Winthrop's promised "city upon a hill" of the early seventeenth century was now a bustling and independent republic, then why shy away from the historic opportunity of letting the world take a tour of its reincarnation as a "free" nation that was "anybody's equal"?[4]

Within a century after Morse's *Geography* was published, American geography had fully transformed into a worldly science and a literally down-to-earth discipline. Professional geographers, including university professors and members of the American Geographical Society and the United States Geological Survey, initiated efforts in the 1880s and 1890s to break free from common nineteenth-century associations of the practice of geography with adventurous masculinity.[5] As Susan Schulten points out, these efforts made it possible for American geography to, at least partly, "emancipate from its thoroughly popular and amateur reputation as a subject concerned with names, places, and facts."[6] By the 1890s, and in step with the nation's ongoing advances into not only Native American land but also the Caribbean and the Pacific, American geography was developing a new conception of the world and of the nation's place within it. This involved both reworking the frame within which it studied the world and complicating the scale on which American geography imagined the United States in its expanse. Reborn as a modern science, American geography was now a most effective tool to present and project the nation's ever-growing political and commercial entanglements with the world at large. Nonetheless, as detailed in the following pages, this rescaling and reframing had its roots in over a century of gradual but lasting shifts in American geographers' worldviews that departed from a revolutionary (localized) assignation of geography to arrive at a patriotic (national) and, later, an international (global) mode of understanding and teaching their subject.

This change in perception was the outcome of efforts to respond differently to old questions, such as where in the world America stood and what center this world had. In the wake of renewed exceptionalist and nativist sentiments among certain strata of American society since the early 2000s, these questions have come to resonate with calls in transnational American studies for scholarship that deconstructs the United States as a "national project" and an exceptional empire.[7] In recent years, the ensuing debates—summed up in works that raise variations of the critical question "How far is America from here?"—have attempted to critically refashion American studies into a discipline that examines the United States as a *trans*national "subject of investigation."[8] Informed by the urgencies addressed by the spatial turn, transnational American studies has called for new modes of mapping the United States, modes that deconstruct the continental territoriality and spatial singularity of our discipline's subject matter by looking at it in macro as well as micro contexts—that is, on local and regional, hemispheric and bicontinental, and global scales.[9] This is a judicious move in writing the history of any nation because, as Thomas Bender asserts in the case of the United States, "the nation cannot be its own historical context."[10] In her critique of national historiography, Angelika Epple maintains that it is "difficult if not impossible to legitimize a presumably given entity ... with fixed borders," even when that spatial entity, say, a nation-state, insists on it. Epple makes it clear that as long as cross-border movements, including the transfer of ideas, objects, and individuals, make borders porous, nation-states or any such macro-level spatial entity cannot be studied as an insulated, self-sufficient, and homogeneous container.[11]

Even so, to examine the geopolitically shifting contours of the United States away from the epistemological frame of the "national," we need to continue raising questions about nationalness and nationalism. Put differently, to problematize the national framework does not mean to suspend the study of the nation—with all the tacit and explicit elements that define or defy it—but "to historicize and clarify its meaning."[12] The discussions in this chapter are consequently founded on the premise that to ask and answer "How far is America from the world?" in an effective manner hinges on attending to a priori, subjacent cultural geographical questions about the rise of the United States to empire and the pluriform encounters of its citizens with others in the world at large. In effect, to ask *how far* America was, or is, from the world necessitates first considering *where* in the world it has been imagined, a question that demands that we examine how Americans and non-Americans have cognitively mapped that world in the first place.

As I make the case in the following pages about the turn of the twentieth century, before Americans could place their country on the world map or measure their distance from its different parts, they needed to closely study a world that had been already explored, surveyed, mapped, and colonized. Consequently, in investigating the zeitgeist almost a century before Americanists attempted to question the predominantly national frame of research on the United States by asking "How far is America from here?" the question should be: "Where *was* America at the turn of the twentieth century?" In responding to this question today and in order to survey the ways Americans responded to it, I explore geography textbooks as a prime example of geographically scripted material for children and as texts that reflect the range of the foundational revisions that took place in Americans' perceptions of the world at large. This investigation lays bare the ways generations of Americans, child and adult alike, experienced an evolution in their "planetary consciousness"—that is, in their investment in interrogating "themselves and their relations to the rest of the globe."[13]

To this end, I first outline a number of ways, both ideological and pragmatic, in which American geography primers evolved over the course of the century, starting with early nineteenth-century patriotic volumes that intended to write the young republic as one unified nation and ending with books that presented geography as a modern science that was instrumental not only in the study of U.S. territoriality but indeed in understanding the whole world. In this part, I investigate Americans' turn-of-the-century self-positioning in geography books and on the newly engraved school maps of the 1890s, tracing the gradual shifts in continual redraftings of a world that had a detachable, movable center. Ultimately, I explore the many contours of late nineteenth-century U.S. geography education around new readings of the concept of "home geography" on a scale far more comprehensive than what it had originally entailed in its early nineteenth-century iterations.[14] An innovative pedagogical method that focused on fostering local spaces, observations, and encounters in order to convey a sense of the world as a grid or a symbiosis of numerous spatial units or "homes," home geography invited students to understand the United States not as a spatially isolated nation but as a home that was comparable to, but also most dear among, many across the globe.[15]

The United States and the World: Revolution in School Geography

Next to Jedidiah Morse's geography textbook, Nathaniel Dwight's *Short but Comprehensive System of the Geography of the World* (1795) was an American geography schoolbook that was published almost two decades after the Revolution.[16] In a series of simple geographical statements, Dwight's *Geography* included long, carefully hand-picked lists of African, Asian, and European cities and specified their distance not from any major American cities like Philadelphia but from London:

- "Ispahan stands 2,460 miles east of London."[17]
- Nova Scotia "is situated between 43 and 49 degrees of north latitude, and between 50 and 70 degrees of west longitude from London."[18]
- "Pekin [sic] the capital of China stands eight thousand and sixty-two miles south-easterly of London."[19]

As it turns out, Dwight's volume presented a half-hearted response to Morse's call for composing *American* textbooks. Smacking of subdued royalist sentiments, it presented young Americans with a world whose center continued to be London—the long-standing seat of the very empire the North American colonists had revolted against two decades earlier. Curiously unpatriotic, it seems that to Dwight and many of his contemporaries who were now citizens of an independent republic, the center of the world remained outside the United States and across the Atlantic.

Even so, as Brückner observes in his study of the fascination with and necessity of revising and reviving geographic pedagogy in the early republic, exemplified by the publication and proliferation of patriotic volumes such as Webster's *Grammatical Institute of the English Language* (1783) and Morse's *American Geography* (1789), "Americans invented their variant of modern nationalism."[20] This nationalism, Brückner further asserts, "evolved out of the ideological tension between regional diversity and geopolitical unity, which Americans were able to reconcile through the programmatic diffusion of geographic literatures that introduced the nation as a material and inherently readable form."[21] Indeed, during the first few decades after the Revolution, schoolbooks proved to be among the main sites where a nationalizing body of "readable" knowledge was being produced for the youth of the new republic across scattered, on-the-move, and newly established communities. Therefore, following the advice of Abigail Adams, Webster, and Morse, proponents of geographical study began to seek a firm place for geography as a "national" school subject, mainly because geography had the capacity not only to bind the expanding nation through cultural, spatial, and political bonds, but also to record and explain its incessant territorial alterations as it expanded westward.

For obvious and urgent reasons, and for a major part of the century American geography had to be continually revised in content, language, and focus. The expansionist policies of the nation, in particular, demanded that American geography be revised and the nation's maps redrawn after each new negotiation or military engagement with Mexico, England, or Russia, and of course following each breached treaty with Native Americans.[22] The locus classicus of the ways geographic revisionism dovetailed with territorial expansionism is the subject of the essay that the populist columnist John L. O'Sullivan wrote in 1845 on the annexation of Texas. In "Annexation," O'Sullivan posited that Texas "is no longer to us a mere geographical space—a certain combination of coast, plain, mountain, valley, forest and stream. She," O'Sullivan opined, "is no longer to us a mere country on the map. She comes within the dear and sacred designation of Our Country."[23] As the addressee of these revisions, the young white children of this simultaneously parochial and expansionist United States who could attend common schools, who had governesses or literate parents at home, or who attended private "academies" established by self-educated relatives or members of their communities had to commit endless geographic facts to memory. This consisted of memorizing a litany of place names; stereotyped, pseudo-historical accounts of the people who lived in various parts of the world; and facts and figures about landscape, agricultural products, and ethnic diversity across the globe. Memorization was facilitated by geographic atlases that were generally long and not entirely free from inaccuracies.[24] These atlases remained in use until almost the middle of the nineteenth century, when new schoolbooks began to be published in a larger (six by nine inches or nine by eleven inches) folio format.[25]

The folio format offered enough space for monochrome illustrations of landscapes, native inhabitants, and wildlife in near and far corners of the world, as well as simple charts that compared the heights of mountains or famous buildings around the globe. In addition, the larger folios included empty maps called "outline maps." Outline maps were used to facilitate step-by-step map drawing and memorization and to prepare students for geography exams.[26] As the century wore on, generations of young owners of geography textbooks would fill their outline maps with an ever-expanding catalog of toponyms that European explorers and colonial officers were busy assigning to newly mapped territories. These empty maps not only served to assist young children in associating toponyms with their corresponding place on the map but served other purposes as well, such as providing young students with an opportunity to assume the role of adult explorers and to vicariously "discover" and reclaim the already discovered, explored, and claimed terra incognita: to spot, name, and tame the spatially unknown.[27]

Even as American educators no longer believed in the urgency of their mission to help white American children in emulating a strong, independent national identity, the patriotic and parochial language of geography readers' prefaces endured long into the century. The "revolutionary" rhetoric of the textbooks waned over time, but the publishing houses' growing commercial interests in a competitive market meant that they would still try to convince the public that it was patriotic to buy American geography textbooks. Marketed as educational material for children and yet entirely dependent on the decisions of adults—including parents, schoolteachers, and district superintendents—a great majority of antebellum geography textbooks were long and tedious, consisting of exceptionally long lists of place names that young children were instructed to memorize. Examples include Daniel Adams's *Modern Geography* and John J. Clute's *The School Geography* (each over 300 pages long), and the slightly shorter, strikingly political volume *First Geography for Children* by Harriet Beecher Stowe (which consisted of 224 pages).[28] Before long, new works, shorter and less overtly patriotic, entered the market. These volumes

FIGURE 1.1 "Going to tell about Geography," Samuel Goodrich, *Peter Parley's Method of Telling about Geography*, 1830 (courtesy of The Newberry Library, Chicago).

adopted a livelier tone and included several visual elements (figure 1.1).[29] One such geography reader, published in several editions from the 1820s to the 1850s, promised to teach geography in a colloquial manner. Compiled by writer, publisher, local politician, and diplomat Samuel G. Goodrich, better known as Peter Parley, the small-format *Peter Parley's Method of Telling about Geography* adopted a simple, playful, but didactic tone throughout. Though patriotic, *Peter Parley's* did not put a disproportionate emphasis on the young republic; rather, it paid almost equal attention to the rest of the known world as it did to the states, landscape, and history of the United States.[30] Goodrich's volume consisted of short lessons in geography, morals, lists of place names, drawings of typical animals on different continents, and history—including myths, hearsay, and stereotypes—about people from various parts of the world.

As Goodrich maintained in the volume's preface, geography suited young children's mental capacities because "geography, more than almost any other youthful study, deals in visible images."[31] It was hardly possible to include sophisticated colored maps in the small schoolbooks of the time; therefore, Goodrich's volume featured simple illustrations of different groups of people, including an illustration that showed supposedly typical wealthy white American citizens. In these illustrations, the volume portrayed people, usually a young man and a woman, busy with their favorite pastimes or stereotypical occupations and in their most representative regional or ethnic costumes. As an exception, some groups of people, such as "Indians" and "Eskimos," were displayed in large groups consisting of adults and children standing by rather idly—an image that placed emphasis on their curiously premodern, communal order of life.[32] Turning its attention to the rest of the world, Goodrich's primer included a long list of people from various nations and colonies. His primer indexed the world via a litany of impressionistic assessments of culture and character—assumptions that were formed at the intersection of his times' racist and colonial views of the world. Giving children a brief glimpse of the Far East, the book applauded the Japanese, who, Goodrich asserted, were "said to be intelligent."[33] Next, Goodrich castigated Hindus, who, he claimed, "sometimes drown their children," and cruel Arabians, who "very often rob those who travel among them."[34] The list went on to mention Russians whose emperor "banish[es] persons to Siberia, who displease him"; the "industrious and happy" Germans, Austrians, and Prussians who "resemble each other in their character and customs"; and South Americans, who, the narrator claimed, "are principally Spanish and Portuguese, and the descendants of Spanish and Portuguese."[35] A few pages later, and looking back at the U.S. involvement in the First Barbary War of 1801–5, he wrote about North Africa's Barbary, "little better than pirates," who "make slaves of those they take in war, or shut them up in prison."[36] As Parley's exhaustive list showcases, the book taught geography as a means to imagine humankind as defined and differentiated by habitat, race, and climate, while it further addressed the relationships individual communities had developed to the land where they lived and to other human beings—as well as, in the case of the Chinese and Norwegians, to animals (figures 1.2 and 1.3). Put differently, the reader was meant to interfuse the spatial regimentation of the world with an evidently imposed colonial order.

For most of the nineteenth century, American geography textbooks imagined the world as ordered into two mutually exclusive categories: the United States as a bounded yet "internally" expanding territorial nation-state versus the non-American world as an expanse of external spaces, whether proximate or distant. By focusing almost exclusively on national territory in to eat with those of another. They are very ignorant, and sometimes drown their children, thinking that they please God by doing so. Some benevolent men, called missionaries, have been sent among the Hindoos to teach them Christianity.

9. India beyond the Ganges includes several countries; as the Birman Empire, Siam, Malacca, and others. The Birman Empire is extensive, and the people are lively, intelligent, and interesting. They are said to be fond of poetry and music. Christian missionaries have also been sent among the Birmese.

Malays.

A Chinese selling Rats and Puppies for pies.

10. The Siamese are still more enlightened than the Birmese. The people of Malacca, called Malays, are wicked, cruel, and ferocious.

FIGURE 1.2
"A Chinese selling Rats and Puppies for pies," Samuel Goodrich, *Peter Parley's Method of Telling about Geography*, 1830 (courtesy of The Newberry Library, Chicago).

13. Denmark and Sweden are cold countries, and not so pleasant as some other parts of Europe. They are not fruitful, but the people are intelligent and generally happy. The people of Denmark are called Danes; those of Sweden are called Swedes.

Swedes.

Norwegian.

14. Norway is a very cold and mountainous country, and the people live principally by hunting and fishing. Bears are very numerous in Norway, and here is a picture of a Norwegian killing a bear.

15. Russia is a vast country, but it is neither a beautiful nor a fruitful country. The people of Russia are generally ignorant, and many of them poor and unhappy.

FIGURE 1.3
"Norwegian," Samuel Goodrich, *Peter Parley's Method of Telling about Geography*, 1830 (courtesy The Newberry Library, Chicago).

isolation and almost entirely uninterested in flagging the United States on world maps, geography textbooks reproduced common views of the time regarding the significance of answering variations on the urgent question, What is this new nation like in geographical terms? Put differently, school geography invited American children to study the United States as a world unto itself and separate from what it understood as the world at large. Consequently, until about the final quarter of the century, the exceptionalist imposition of this split imaginary of the world foreclosed the possibility to write the United States into the common geographic narratives of the time. If not elevating the United States above geography, this narrative most certainly corroborated the dominant rhetoric of the times that insisted on Americans' parochial disinterest in, and alleged reluctance to participate in, the goings on of the rest of the world.

The United States in the World: Evolution in School Maps

In this metageographical disjunctive worldview, and while the United States viewed itself as its own center, it conceded that the rest of the world had another center. As late as the 1850s, projections of the earth in school geography books reproduced the common Eurocentric world projections that dominated the cartographic imaginations of the world—those in which Europe, and inevitably Africa, occupied center stage. A concomitant but minor shift was also visible in projections that tilted slightly to the west, centering the world on the Atlantic. This latter move helped to maximize focus on the United States, bolstering its readiness to admit its growing ties to other parts of the world, primarily to Europe. Over time, with a gradual shift in desire to emphasize the bonds between the nation and the world in the final decades of the century, and as Americans developed a different understanding of their place as a nation among many, several American geography primers and readers began to remap the world in attempts to put the United States at its center. To better understand this shift in scale (from national to global) and perspective (from nationalist sentiments to imperial ones) at the service of the changing geopolitical urgencies of the nation, I spend the next few pages interrogating the twofold transformations in geography textbooks—namely, changes that took place in astronomical geography and, only later, in physical geography.

With roots in the early revolutionary years, aggressively nationalist features of school geography readers and primers throughout the second half of the

FIGURE 1.4
"Comparative size and appearance of the Planets," William Channing Woodbridge, *Peter Parley's Method of Telling about Geography*, 1838 (courtesy of Special Collections, Monroe C. Gutman Library, Harvard Graduate School of Education).

century, including the gradual shift that resulted in centering the world on the Americas, did not begin to appear in the subdiscipline of physical geography. Rather, it started in what was then called "astronomical geography."[37] Imitative of common school geographies published in Europe (especially in Britain) and aspiring to be as scientific as possible, early American geographies included information not only about planet Earth but also about other known planets in the solar system (figure 1.4). Later in the century, as more facts came to be known about other planets and their satellites and as more efficient techniques began to be implemented in the United States for print-

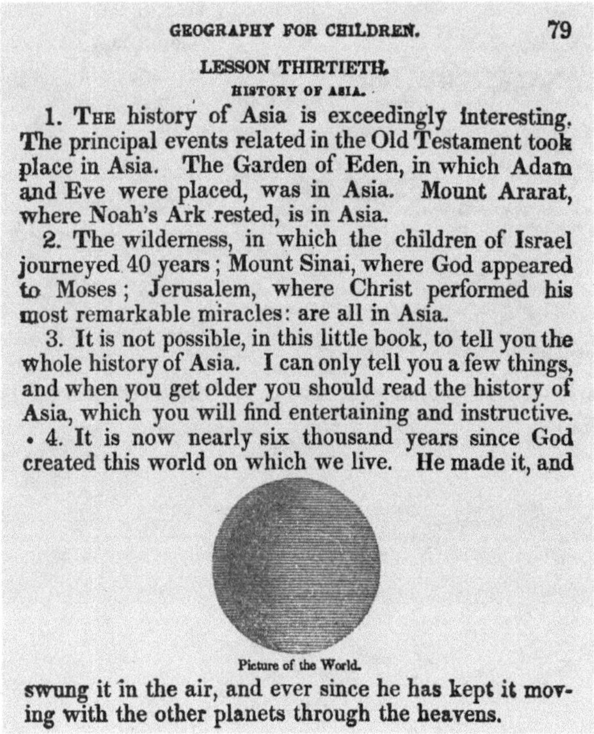

> GEOGRAPHY FOR CHILDREN. 79
> LESSON THIRTIETH.
> HISTORY OF ASIA.
> 1. THE history of Asia is exceedingly interesting. The principal events related in the Old Testament took place in Asia. The Garden of Eden, in which Adam and Eve were placed, was in Asia. Mount Ararat, where Noah's Ark rested, is in Asia.
> 2. The wilderness, in which the children of Israel journeyed 40 years; Mount Sinai, where God appeared to Moses; Jerusalem, where Christ performed his most remarkable miracles: are all in Asia.
> 3. It is not possible, in this little book, to tell you the whole history of Asia. I can only tell you a few things, and when you get older you should read the history of Asia, which you will find entertaining and instructive.
> • 4. It is now nearly six thousand years since God created this world on which we live. He made it, and
>
> Picture of the World.
>
> swung it in the air, and ever since he has kept it moving with the other planets through the heavens.

FIGURE 1.5 "Picture of the World," Samuel Goodrich, *Peter Parley's Method of Telling about Geography*, 1830 (Division of Rare and Manuscript Collections, Cornell University Library).

ing illustrations in larger folio format, simple illustrations of Earth and other celestial bodies became standard in such textbooks as Woodbridge and Willard's *A System of Universal Geography* (first published in 1824).

In this and other similar geography textbooks of the time, the focus was mainly on transmitting cutting-edge information about observing and measuring celestial bodies and about technological advances in manufacturing astronomical instruments, at the forefront of which stood leading European astronomers such as Friedrich Bessel (1784–1846), Sir John Herschel (1792–1871), Frederico Augusto Oom (1830–90), and Hervé Faye (1814–1902).[38] Authors of these nationalistic volumes made efforts to market astronomical geography as an up-to-date scientific field and to keep the American public, including American children, informed about the dynamics of astronomical geography as a popular European science. An illustration in *Peter Parley's Method of Telling about Geography*'s eighth edition (1838), for example, compares the size of different planets, depicting Earth as a small globe but without rendering any landmasses visible on its surface (figure 1.5).[39] In time, the emphasis on new scientific developments would be adapted to American physical geography as well.

Though not all planetary charts depicted Earth's continents, let alone national boundaries, American astronomical geography primers gradually came to depict the United States as the brightly lit face of the earth that was swung abreast of the sun while the planet's other landmasses were rendered invisible. To create this effect, little change needed to be made to the Mercator projection that had been, for well over three centuries, the most widespread visual representation of the world, with Europe at its center. The shift to an American center is traceable, however, not only in new volumes compiled throughout the century, but even across multiple editions of the same volume. Two identically titled "Picture of the World" illustrations of planet Earth in Samuel Goodrich's 1830 and 1838 editions of *Peter Parley's Method of Telling about Geography*, for instance, demonstrate how the globular depiction of the world, and the position of the young republic in it, had changed within a single decade in antebellum America. On the one hand, the 1830 edition includes a North Pole projection of the planet where the link is visible, perhaps even emphasized, between the various landmasses of North America, Asia, and Europe on a star-clad earth—a step between centering Europe and centering America, a moment at which the desire to decenter Europe had arrived but the impulse to center the United States not fully materialized. As depicted in this illustration, the world centered on a thinly populated node, the North Pole, that offered a "land-hemisphere" view of the planet. Asia stood on top of the globe, while the North American continent was placed on the lower end of the globe's visible face—locations that defy the common north–south axis in world cartography that the modern eye so quickly conceives as the "norm."

On the other hand, the illustration of the same title in the 1838 edition zooms slightly out to include the sun and the moon. In this version, the author decided to place the north at the top of the illustration and opted to center the globe on the Atlantic Ocean, perhaps in order to create a distancing effect between the rhetorically isolationist United States and the rest of the world (figure 1.6). Additionally, by including the sun in the illustration, the 1838 picture of the world introduced the notion of time hand in hand with the notion of space in astronomical geography. While Asia and the entirely invisible Australia experienced the night's dark and resided supposedly in the past, western Europe, North America, and the northern parts of Africa and South America were brightly lit by the sun, locating them in the present. It was no accident that in this visual metaphor for relative civilizational devel-

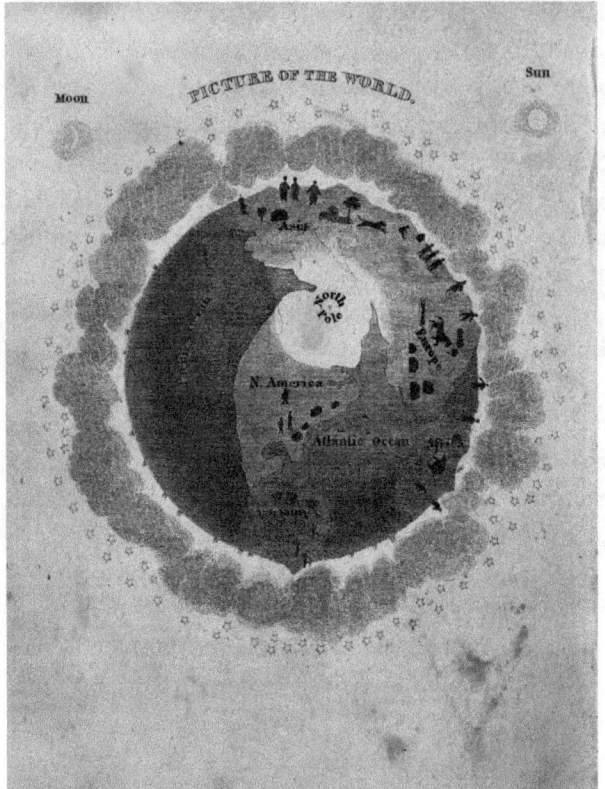

FIGURE 1.6
"Picture of the World," Samuel Goodrich, *Peter Parley's Method of Telling about Geography*, 1838 (courtesy of The Newberry Library, Chicago).

opment, the parts of the world to which the white race was more or less unanimously believed to have spread are the same areas of the planet that Samuel Goodrich depicted as beneficiaries of the sun's warmth and light. (Compare this to a late nineteenth-century map showing the distribution of "races of men," shown in figure 1.7.)[40]

Inclusive of maps that expanded on the parochial notion of a "city upon a hill" to depict the United States as the land under the sun, astronomical geography gives a measure of the lengths to which American geographers were ready to go in order to place the United States at the center. In sum, by century's end, shifts such as these—the visual replacement of the northern continental connection stressing Europe's dominance with a northern Atlantic Ocean across which the light of science and reason had spread to the United States—imagined the United States as the center not of the planet but of the entire universe. While some school geography primers continued to

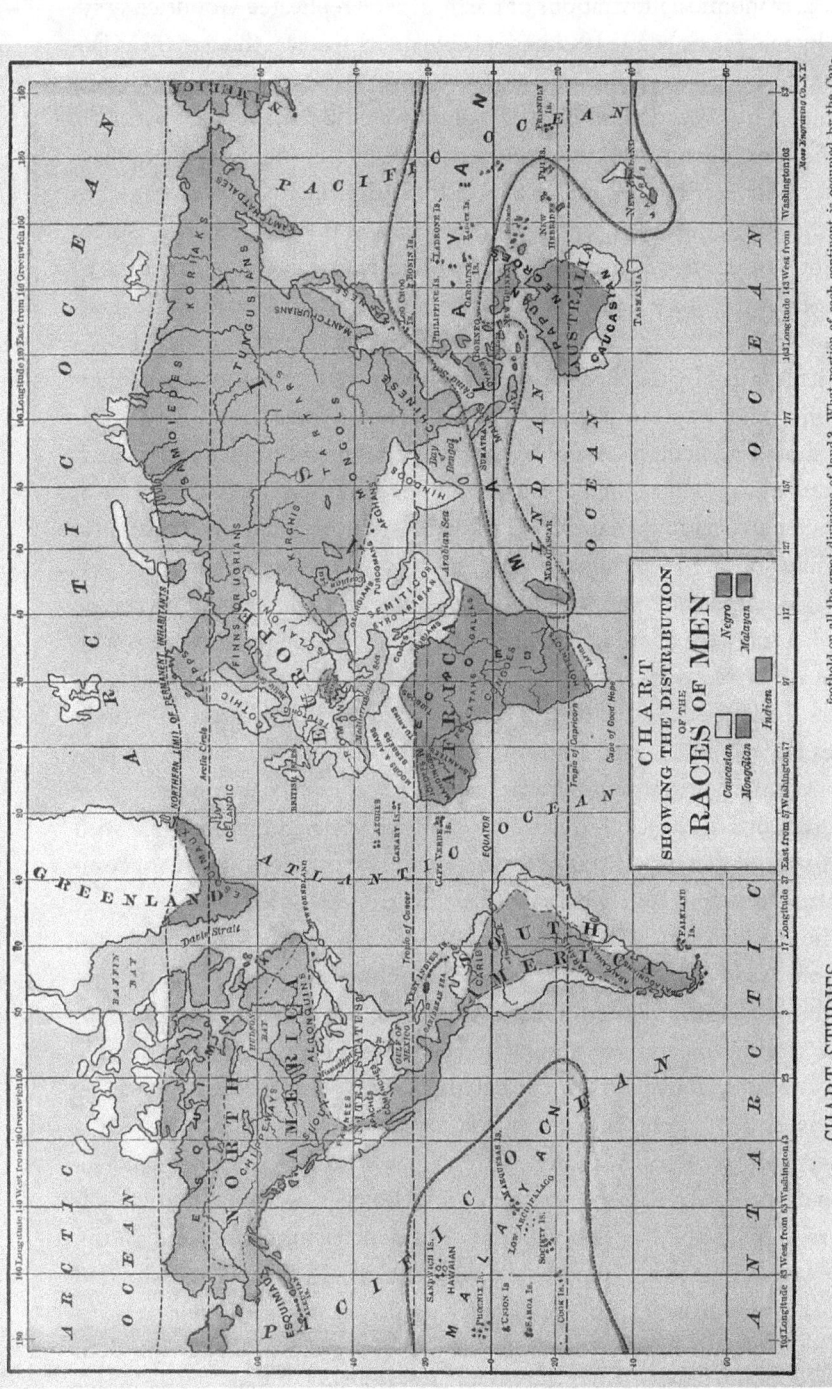

CHART STUDIES.

Which of the five races occupies the largest portion of the earth? Which race is confined principally to islands? Which continent is peopled by the largest number of races? Which race has a foothold on all the great divisions of land? What portion of each continent is occupied by the Caucasian race?* What parallel runs nearest the northern limit of permanent inhabitants?

* The pupil should know that the color on the chart indicates the most numerous, or the leading, race in any region; that is, the race to the exclusion of all others. In the case of Europe, for instance, the color shows the Caucasian to be the dominant race, but we know that there are millions of negroes here, too

FIGURE 1.7 "Chart Showing the Distribution of the Races of Men," Sanford Niles, *The Complete Geography: Mathematical, Physical, Political,* 1889 (courtesy of The Newberry Library, Chicago).

include astronomical illustrations of Earth, several replicated Goodrich's 1838 model by making only one landmass visible on its surface: the Americas facing the sun.

At the same time, astronomical geography as a simultaneously patriotic, scientific, and romantic section of American geography began to matter less and less to American educators. In the postbellum United States, more elaborate and at times more fully colored illustrations of the globe, as well as later full-color world maps, began to push the Atlantic Ocean or the Americas to the center of the images and to the center of their readers' spatial imaginaries as an act of (inter-)national self-awareness. A survey of geography books makes it clear that by the 1870s, Americans had more thoroughly turned their attention from the solar system back to the planet. By this time, they had begun to show a meticulous awareness of other continents, other nations, their inhabitants, and their natural resources. In effect, during the final quarter of the nineteenth century, illustration after illustration and world map after world map appeared in school geographies that, while zooming *in* and rotating the globe to center the world on the young republic as its most crucial point, also zoomed *out* enough to draw attention to the nation's relations with the non-American world. Sarah S. Cornell's 1878 *Intermediate Geography*, for example, includes a Mercator projection of the world (suggestively titled "Routes between the Grand Divisions of the World") that shows the Americas at its center. By that point, it was imperative to replace former, romantic representations of the United States as it stood alone among the stars with more matter-of-fact depictions of the nation's place in a dense tangle of commercial and diplomatic networks on the surface of the earth.

In the 1896 edition of William Swinton's *Grammar-School Geography* (in print from 1880 to 1896), all the physical and political maps of the world center on Europe. However, the volume's "Commercial Chart of the World" centers the world projection on the United States, signifying the considerably greater commercial ties of the nation with the world and its prominence, by the 1890s, as the largest producer and exporter of a wide variety of goods.[41] By the same token, Rand McNally's volume *Primary School Geography* (1894), popular among late nineteenth-century professional geographers and schoolteachers, depicts the North American continent facing its readers on the front cover and the rest of the world on its back cover—except for Africa and Australia, which are not visible at all.[42] Yet another example is *Complete Geography*, by H. Justin Roddy (1902). Published at the dawn of the twentieth century, Roddy's book depicts the world with the United States at its center in a map titled "Map of the World Showing Colonial Possessions, Principal

Commercial Routes and Telegraph Lines" (figure 1.8), thus subordinating European colonial relations—previously the focus of geography primers—to the United States' role in global trade, communications, and territorial acquisitions.[43]

After all, while traditionally Eurocentric world maps placed emphasis on the distance between the United States and the rest of the world, U.S.-centered projections not only insisted on the centrality of the United States in world affairs but also stressed its proximity to other continents to the east and west. In Clif Stratton view, at this time school geography "opened for schoolchildren the widest possible lens through which to see themselves and the United States *in the World*" (emphasis mine).[44] By the turn of the century, this sweeping repositioning trend enabled American cartographers to include the Philippines and Hawai'i as adjacent islands in the country's vicinity. By 1900, the Philippines, for instance, was no longer imagined as a remote archipelago at the eastern edge of Asia—as it had previously appeared on the right-hand margin of the Eurocentric Mercator projection. Rather, it was now a significant U.S. "possession" that neighbored its most recent mother country in the West. No doubt these U.S.-centered world maps were far better suited to justify ambitions to annex the Philippines in the aftermath of the Spanish-Cuban-Philippine-American War for the simple reason that they repositioned the archipelago to the west of the country—a stretch of islands that seemed easily reachable as the "natural" next stop in the westward expansion of the country into the Pacific Ocean. Ultimately, school geography now underlined the prominence and centrality of the United States as an emerging global empire on maps and globes. By 1900, the gap between geography as a science and as a tool of empire was almost completely bridged.

"Home Geography" and "the World Whole": Contours of Modern Geography

Although not as popular as math and English among district superintendents, school principals, and teachers, geography received relatively more attention than history in American school curricula.[45] According to the *Report of the Bureau of Education*, in 1876 (that is, exactly a century after the American Revolution) students in Ohio took American history only in the eighth grade.[46] The report confirmed that in 1880, only 31,171 Ohio students took American history, and even fewer took general history as a school subject. At the same time, however, a striking total of 267,618 students in Ohio were reported to have taken geography.[47] Late nineteenth-century changes in the

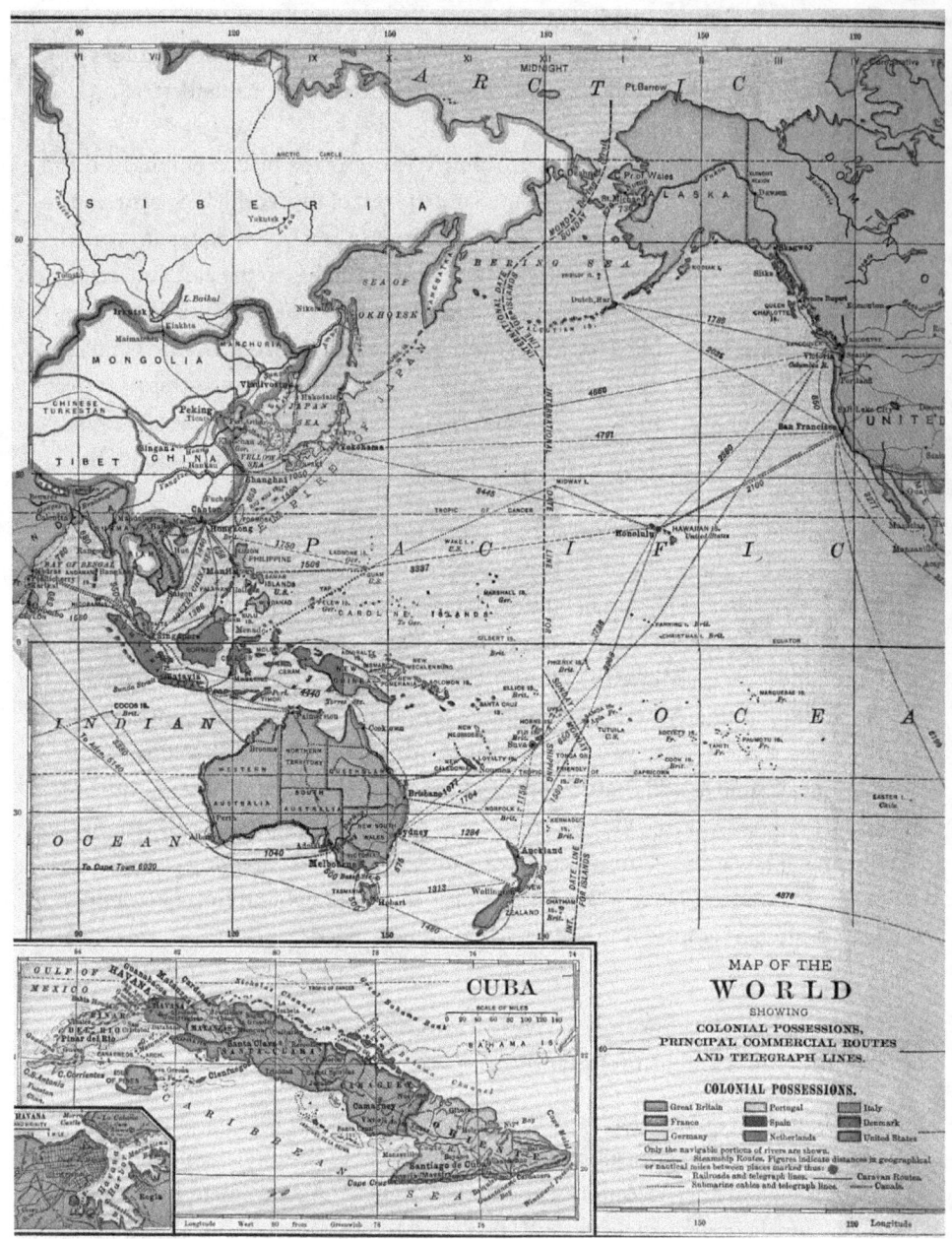

FIGURE 1.8 "Map of the World Showing Colonial Possessions," H. Justin Roddy, *Complete Geography*, 1902 (courtesy of Special Collections, Monroe C. Gutman Library, Harvard Graduate School of Education).

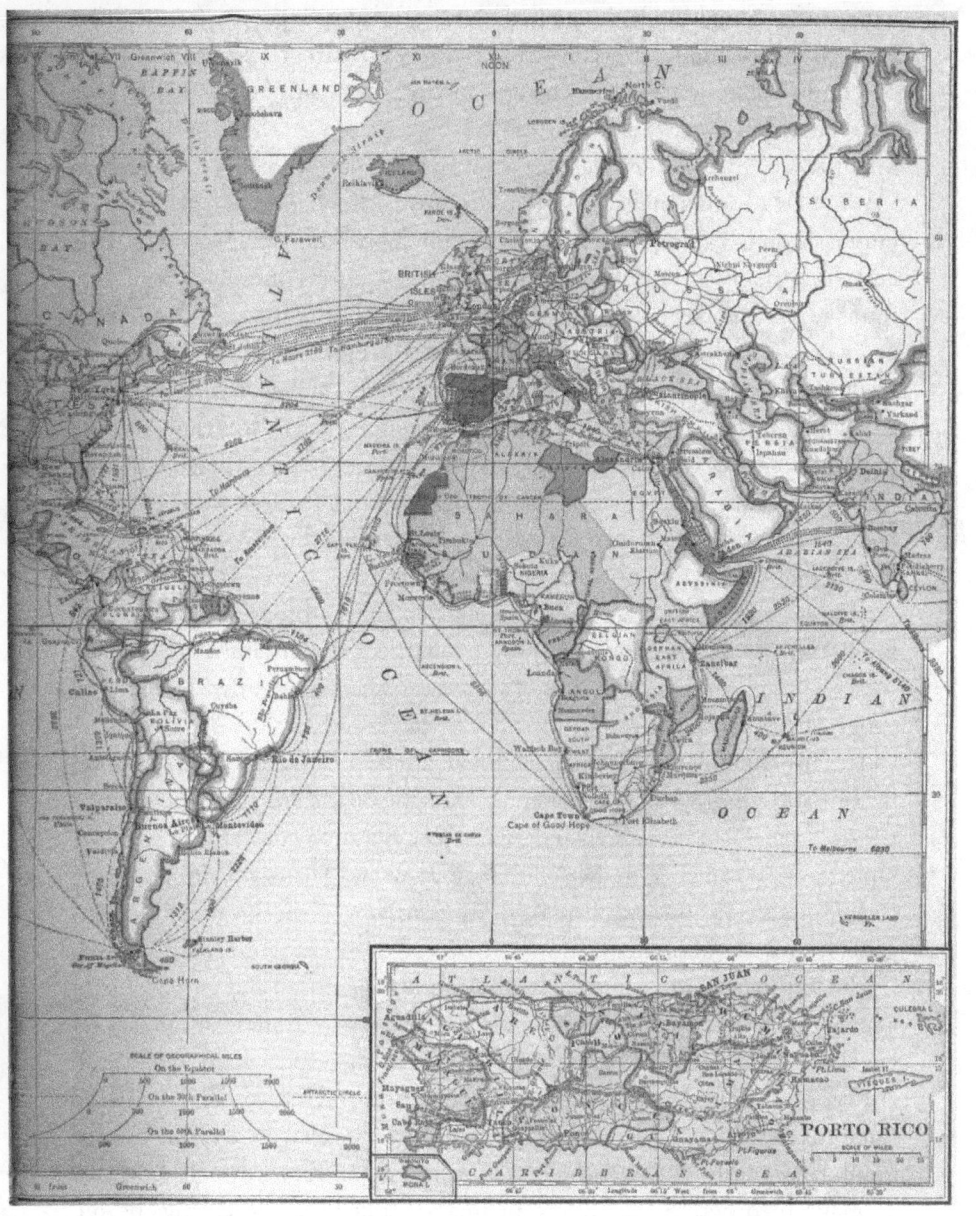

nature of geography, which entailed plans by academic geographers at Harvard and other leading universities and colleges to turn it into a modern academic discipline, resulted in changes in both popular and curricular sites through which Americans were exposed to geography. At the curricular level, for instance, the Committee on Secondary School Studies, established in 1893 as part of a national campaign to improve school curricula, placed extra emphasis on geography and its potential for subsuming other sciences. This and other national reform committees formed during the 1890s and early 1900s under the auspices of the National Education Association called for change, first in the foundational definition of geography as a science, to include the interaction of humans and environment, and second in the ways geography was taught across the nation. Geography had by this time become so popular that by the second half of the 1890s, no less than 90 percent of high schools in Pennsylvania included geography as a mandatory course.[48] Such a rise in numbers had its reasons mainly outside the schoolroom, in external factors that affected school geography as a discipline both directly and indirectly. Geography's human-driven objectives and anthropocentric content, for instance, were emphasized by professional geographers, university professors, textbook authors, and teachers whose theories undergirded and defined geography as a discipline.

Despite their identification as different subjects in the report cited above, history and geography were generally thought of and taught together. The *Central School Journal*, for instance, listed one hundred questions in its geography section called "Questions in Geography and History."[49] Proposed by B. A. Hinsdale, an educator from Ann Arbor, Michigan, the questions blended geography with alternative history scenarios. Question 55, for instance, asked, "If the Europeans who came to North America in the 16th and 17th centuries had landed on the Pacific Slope, would the progress of American civilization have been more or less rapid, and why?"[50] Another question asked, "How have rivers of the United States influenced emigration and settlements?"[51] By the same token, believing that "the growth of history is largely a result of geographical conditions," *A Pathfinder in American History* advocated the use of maps in order to "go sailing in imagination with the daring, heroic discoverer"—that is, when studying the historical voyages of Christopher Columbus.[52] This and many similar books also advocated the parallel study of geography and history mainly because they believed that geography had the power "to furnish a stage for the great drama of human history."[53]

In 1870, geography began to be listed as an admission prerequisite at the two leading private and public American universities, namely Harvard Uni-

FIGURE 1.9 "Programme for Fully Graded System," Edwin Grant Dexter, *A History of Education in the United States*, 1904 (courtesy of Special Collections, Monroe C. Gutman Library, Harvard Graduate School of Education).

162   *The Growth of the People's Schools*

PROGRAMME FOR FULLY GRADED SYSTEM

Branches	1st year	2d year	3d year	4th year	5th year	6th year	7th year	8th year
Reading	10 lessons a week			5 lessons a week				
Writing	10 lessons a week	5 lessons a week	3 lessons a week					
Spelling Lists				4 lessons a week				
English Grammar		Oral, with composition lessons				5 lessons a week with text-book		
Latin								5 lessons
Arithmetic	Oral, 60 minutes a week	5 lessons a week with text-book						
Algebra							5 lessons a week	
Geography	Oral, 60 minutes a week	5 lessons a week with text-book *				3 lessons a week		
Natural Science + Hygiene	Sixty minutes a week							
U.S. History							5 lessons a week	
U.S. Constitution								5 les.
General History	Oral, sixty minutes a week							
Physical Culture	Sixty minutes a week							
Vocal Music	Sixty minutes a week divided into 4 lessons							
Drawing	Sixty minutes a week							
Manual Training or Sewing + Cookery							One-half day each	
No. of Lessons	20 + 7 daily exer.	20 + 7 daily exer.	20 + 5 daily exer.	24 + 5 daily exer.	27 + 5 daily exer.	27 + 5 daily exer.	23 + 6 daily exer.	23 + 6 daily exer.
Total Hours of Recitations	12	12	12½	13	16¼	16¼	17¼	17¼
Length of Recitations	15 min.	15 min.	20 min.	20 min.	25 min.	25 min.	30 min.	30 min.

* Begins in second half year.

versity and the University of Michigan.[54] As other universities followed suit, an increasing number of high school students who planned, and could afford, to enter college needed a good grasp of the discipline—a change that, in turn, led to increasing standardization of secondary school geography teaching.[55] In the meantime, geography as a school subject had gradually departed from its beginnings as a conglomerate discipline taught by ministers or geologists. In fact, it had begun to adopt the scientific language of evolution. By the 1890s, it had started to develop into a science that sought answers to ever more pressing questions regarding the material complexities of human life on earth. Reports from the Committee on Elementary Education suggest that by 1900, geography was taught all the way from the first to the eighth grade, orally in the first two grades and later through lessons taught with textbooks.[56] As figure 1.9 confirms, geography occupied a larger proportion of the students' weekly schedule than did history or even arithmetic.[57] Only the three basic subjects (English, reading, and writing), the table confirms, competed in frequency with the rate at which geography was taught at American schools.[58]

By 1900, specialized journals of pedagogy in geography, such as the *Journal of School Geography*, began to appear, too. Edited by Richard Elwood Dodge—a professor of geography at Teachers College, Columbia University, and author of several geography primers—the journal advised schoolteachers on proper classroom teaching methods. The journal further labored to unify geographical education in far-off corners of the nation and to provide lists of questions in order to standardize geography examination across the country.[59] As geography confirmed its place in school curricula, larger numbers of American children were now exposed to geography through media other than the adventurous tales of exploration that they avidly read in works of fiction or popular travel literature, such as Thomas Wallace Knox's Boy Travellers series.[60] Modern geography curricula would start with local and national geography as a preparatory step to world geography. In this frame, once children learned to observe and study the geography of the places where they lived, and after they learned about U.S. geography, they would get ready to fashion ideas, mainly in the form of comparison and generalization, about how other parts of the world were similar to or different from American geography.[61]

Refashioned from an early nineteenth-century locally focused approach to the study of geography, which had been developed by Emma Willard and William C. Woodbridge, this approach to the simultaneous and comparative study of American and world geography was called "home geography." Invoking but also expanding the original scope of the concept, William Morris Davis (1850–1934) was a member of the first generation of professional American geographers who called for sweeping changes to be made to the methodology with which geography was taught. As he explained in 1897, Davis had molded a draft of home geography that was anything but parochial: "The study of home geography does not find its chief recommendation in the local information that it provides, but rather in the aid that it furnishes through local examples to the general study of geography, by giving full meaning and reality to geographical facts and relationships the world over. The reason for this," he continued, "is that geography as a whole is hardly more than a compilation of innumerable local or home geographies."[62] Published in the beginning pages of the first issue of the *Journal of School Geography*, Davis's article communicated the significance that the method (and the worldview it replicated) had for the pedagogical agenda of turn-of-the-century American geography.

A professor at Harvard, a member of the geography committee on the Committee of Ten, and later dean of the Association of American Geographers, W. M. Davis was a prominent figure in the generation of American geologists and geographers who were most influential in modernizing American geography at the turn of the twentieth century. In his writings, Davis advised schoolteachers to consider elementary school geography as a collection of lessons in home geography. In Davis's view, co-opting an old notion for the study of world geography at a globalizing moment—or, in his own words, "the extension of the local to the remote"—was instrumental in transforming the examination of local, national, and world geography.[63] Despite advocating for the new, more methodical draft of home geography, Davis did not fail to caution about rash generalizations with regard to world geography: "It is not only through the principle of likeness," Davis maintained, "that home facts are useful in describing distant facts; the principle of contrast is no less helpful and important."[64] In effect, mindful of the racial, social, and civilizational hierarchies that many Westerners believed in, Davis's model of elementary geographic instruction underlined the similarities in various local geographies as much as it heeded to the differences in drawing the bigger picture.[65]

Closely echoing Davis, Richard Elwood Dodge—editor of the *Journal of School Geography* and author of numerous popular school geography books—prefaced his *Home Geography and World Relations* (in print from 1904 to 1921) with similar views on the affinity of national and world geographies.[66] Targeting primary school children, the first half of Dodge's *Home Geography* was dedicated to home and the second half to world geography. Though intended as a tool to make the study of geography personal and pleasant for young children, Dodge's notion of home geography was heavily invested in the tangle of commercial reciprocities and political exchanges between different nations:

> The Home Geography is brought to a climax in a few chapters devoted to the relations of each individual to the world as a whole. From such a view point these relations are readily part of Home Geography, because they are personally important to each child and adult. The interdependence of individuals and of nations is thus emphasized and pupils realize that the trade and industrial relations of their home community are to be found repeated all over the world, and that they are also an epitome of national relations. Thus a deeper human significance is given to the text, and pupils are taught to view sympathetically peoples who at first seem very distant in manner of life as well as place.[67]

In line with Davis's emphasis on "likeness" as well as "contrast," Dodge points here to a number of elements that were central to the modern definitions of geography: the extra-scalar study of geography through comparison and contrast; the benefit of turning world geography into palpable, almost intimate, subject matter for the child; and the indisputable impact human beings have on commerce and politics and, in turn, on how polities behave.

With Clara Barbara Kirchwey, Dodge coauthored *The Teaching of Geography in Elementary Schools*. In it, Kirchwey and Dodge divided the subjects that were to be taught in elementary geography into "social units" and "earth units" to emphasize the "organic," or human, element of geography, which—according to the wisdom of the time—was considered the primary reason for the study of geography. In their words, the social units explained "the grouping of people in homes, villages, towns, cities, states, and nations; the needs of communication between groups of people and how they are met; . . . the industrial conditions to be seen in neighboring localities, and how children and their parents are related to the products of different industries."[68] The earth units, on the other hand, touched on "the relations of life—plant, animal, and human—to the surface features in the neighboring locality, to the drainage features, to the weather, to soils, to direction, and to distance and maps."[69] These two "classes" complemented each other in that they gave meaning to the relationship between men and nature right from the interiors of American homes to the outskirts of "the world whole."[70] In the authors' view, the city or the nation—in sum, what they called the political scales of studying geography—does not have the same significance and relevance for the child as do the more intimately conceived "homes" that weave the global together: "Such a method of procedure along lines of mutual relationship is more natural than a procedure along political boundaries. Why should a child be led outward in concentric circles from schoolroom to the school yard, to the street, the block, the ward, the city, the county, the state, the nation, when the intimate things of his life relate him to more distant regions than to these political units of area?"[71] In other words, the study of trade, communications, and the industrial relationships of people to nature and of people to people facilitated learning geography at the elementary level far more effectively than the study of political units and the boundaries between them would.

Once home geography had been thoroughly examined, scales such as the national (and the corresponding boundaries between nations) could be suspended in order to permit young children to form an understanding of world geography as a matrix of comparable, if not necessarily identical, home geographies. At the same time, in adopting "home geography" as a model to rethink the global, Davis did not call for a total reenvisioning of the world as a seamless whole in which empires, sovereign nations, colonies, and protectorates were traded for a litany of equally modern, comfortably habitable, safely portable homes. Nor was his model an attempt to translate the empire into a project of expansion that "imagines a home co-extensive with the entire world."[72] Far from that, while collapsing other hierarchies (between regional, national, and global) to make young children's first encounters with geography less cumbersome, home geography did not simply underscore the subtle hierarchies between homes across the globe; indeed, by placing emphasis on how they differed from one another, it further complicated the picture by positing "home" in a thoroughly racialized binary opposition to whatever it walled off.[73] It is impossible, for example, to imagine that Davis or Dodge would have conceded that Africa, China, or Latin America housed more than a handful of "homes" outside white settlements and colonial and missionary outposts established there by white colonizers. In effect, their draft of "home geography" raises an urgent question: Did children find only a fixed number of homes across the globe, or was "home" a shifting, multiplying quality of space that formerly uncivilized territories could potentially "acquire" through submitting to colonization and missionary work and with the help of infrastructural improvements such as construction of dams, schools, canals, and so on?[74]

While subscribing to the basics of home geography, geographers such as George D. Hubbard and Albert Perry Brigham turned their attention to geography in secondary schools. In an article that appeared in the 1903 volume of *Journal of Geography*, Hubbard, of Cornell University, echoed Dodge and Kirchwey's sentiments on the indispensability of geography in American secondary school curricula. While promoting geography as a key subject to be taught in American high schools, Hubbard put forward categories similar to those of "social" and "earth" geography—namely, "the study of life consequences (static geography)" in their causal relationship with "that of earth causes (dynamic geography)."[75] Hubbard was a keen advocate and writer in the burgeoning field of anthropogeography and had published extensively on physical and human geography. Keeping westward expansion during the course of the century in view, his scholarship involved research on human migration in response to the need to search for natural resources such as precious metals. Hubbard was among the most resourceful turn-of-the-century American geographers who sought to refashion the emerging science of geography into a multifaceted discipline.[76] Under the influence of, and yet also at times in reaction to, coterminous changes occurring in the field of geography in western Europe, Hubbard, Brigham, Dodge, Davis, Kirchwey, and their colleagues constituted a school of professional and college geographers who conceived of modern geography in terms of its political, human, and commercial applications. For instance, Hubbard's colleague Albert Perry Brigham (1855–1932), then professor of geology at Colgate College and later at Harvard, made unequivocal claims as to the evolutionary foundation of human-centered geography: "No man is prepared to think truly who does not, in some measure, appreciate the world and its organisms as evolutionary. We deal with facts, but with facts genetically related, and hence not nakedly informational. We study concrete things, to be sure, but in the light of their history and as having their goal in man. For this point of view the secondary school has no resource but geography."[77]

As editor of the *Bulletin of the American Geographical Society* and author of several commercial and physical geography textbooks, Brigham did not fail to emphasize the necessity, on a national level, for American citizens of relating to the larger world by learning world geography: "Here is a great body of facts about the world in which we have our home. These facts do not merely supply essential knowledge for our daily conduct; they enlarge the mental horizon, add to our dignity and sense of responsibility as citizens and rulers of the world.[78] In Brigham's view, American children were to study the world not only to practice how to rule over it when they grew up but, more importantly, to learn how to treat it as belonging to one and the same domicile—that is, to the very nation in which they were most "at home." Put differently, even as turning into citizens of the world was predicated on growing up (and becoming adults), children could already visualize instances of going far in order to practice ruling over the world while attending geography courses at their local school.

While prominent European geographers such as Paul Vidal de La Blache (1845–1918) pursued "regional geography" (the so-called *Landschaft* geography), their American colleagues adhered to home geography as a means that better corresponded to their new role as citizens of a globalizing empire.[79] Blache looked at French rural communities to develop his ideas and, as a result, produced a body of knowledge that, while appealing outside the perimeters of French geography, would fail to explain the complexities of geographical change in the heavily industrialized and urbanized United States. Therefore, while thinking within more or less parallel frameworks—that is, reliance on close examination of a locality in order to infer general patterns about the wider world—American geographers pursued a different objective. Tracing the multiple exchanges between the parochial and the global, they endorsed home geography as a productive conceptual and methodological framework through which to understand *both* American and world geography. In retrospect, the advantage of the American approach seems to stem primarily from its scalar adjustability. After all, it could be deployed both as a helpful pedagogical tool to facilitate children's spatial grasp of what they associated with the immediate locale in the study of American geography and as a methodological toolbox that helped pupils make sense of the inaccessible global in the study of world geography. Even as the late century recast model of home geography (by Davis and Dodge) failed to spell out the clear terms of similarity and difference that set up homes across the globe in a clear hierarchy, home geography is an essential marker of how differently turn-of-the-century American adults looked at themselves and the world: no more a distant and sequestered geographical entity but rather one prominent locality among many comparable ones, thus writing the child's home into the nation and further into the world.

Effortlessly comprehensible and yet dauntingly versatile, home geography enlarged young Americans' spatial appreciation of "home" as a seemingly contiguous, pristine, uniform national whole that integrated a larger, far less immediate area extending beyond the actual houses, farms, and towns they lived in. In effect, undermining the binarism at work between centers and margins, home geography more than simply vacillated between the local and the global. On the one hand, home geography had the power to reduce the incomprehensible enormity of the world and the relative distance between its various parts and children's homes by teaching them to view other localities as homes that were populated by children like them.[80] American geographers saw home geography as a tool that could be instrumental in acquainting children with the world both as bodies of land and water rich in exploitable resources and as a multitude of markets ready to pay for the finished goods produced by the labor force that was employed and exploited by the empire. Indeed, unlike contemporary ecocritical criticism that celebrates the child as our only hope for survival as a species, home geography imagined the child as the future leader of a contiguously colonized world.[81] On the other hand, imagining the surface of the earth as a catalog of homes, home geography had the power to, at least in theory, eclipse age-old cartographic altercations as to where the world's center was. Immersed in the imperial views of the time, advocates of home geography believed that colonization was headed toward a long and prosperous tomorrow in which today's children were going to lead uninterrupted phases of expansion (of the territorial and ideological span of home) and exploitation (of what resided outside its walls).[82]

Teaching Home Geography: The New Geography Primer

Reproducing revised drafts of the world, modern textbook series such as the school primer *Harper's School Geography* nourished children's geographical appreciation of the world and the various continents with an evident focus on the United States and its commercial interdependence on and rivalry with other "leading nations of the world" (figure 1.10).[83] In solemn scientific language, the series (first copyrighted in 1875) introduced its young audiences to geography as an indispensable science and mapmaking as its accompanying craft, providing facts about peoples on different continents and their respective cultures and customs. The purpose of the series, the publisher noted in its advertisement for the 1886 edition, was to educate children into informed citizens who would, along with adopting other habits of a modern populace, become informed readers of newspapers: "The study of geography is now, much more than at any former period, an essential element in education. It is second in importance only to reading, writing, and rudimentary arithmetic. The newspaper is and must continue to be the chief source of that knowledge of current events which is indispensable to every intelligent person. Its telegrams and other items and articles necessarily assume that the reader possesses a knowledge of certain geographical facts. It is the aim of this work to present and impress these facts."

Understood in Machiavellian terms, it was common practice for geographers in the sixteenth through the eighteenth centuries to be at the service of royalty as explorers, strategists, and cartographers and to offer advice to the prince. Beyond any doubt, by the nineteenth century, geography, especially its turn-of-the-century American variation, addressed an audience far more inclusive than politicians and the elite. The goal of American geographers, and by extension of school geography, was indeed to fulfill white American citizens' need as well as desire for a dependable geographic knowledge of the world before they could make any sense of what was going on in it.

The Guyot Geographical Reader and Primer: A Series of Journeys Round the World, which was in print from 1882 until long after the death of its original author, the Swiss immigrant Arnold Henry Guyot (1807–1884), further exemplifies this approach to geography. Divided into two parts, *The Guyot* introduced geography in simple language. Geography, Smith and Guyot write, is "a description of the earth."[84] According to the preface, the reason to teach geography is that no matter how carefully people use their eyes, they still cannot see the entire world and they cannot trust that the world is round unless they trust the hard work of great men—the geographically privileged indi-

FIGURE 1.10 "Geography is the description of the surface of the earth, and its countries and their inhabitants," *Harper's School Geography, with Maps and Illustrations*, 1886 (courtesy of Special Collections, Monroe C. Gutman Library, Harvard Graduate School of Education).

viduals referred to in the introduction—throughout centuries of discovery and exploration. First composed in the 1880s at a time when nativist sentiments opposing new waves of immigration from southern Europe and East Asia were at a record high, the preface to *The Guyot* maintained a forbidding us-versus-them rhetoric as it introduced white American children to the world: "Geography teaches us . . . about interesting countries in far-off parts of the world; about strange people, who look very unlike us; and about strong and fierce animals and curious plants, which we have never seen, and which could not live in a country like ours."[85]

With a more or less indifferent tone throughout, the reader part of the book includes illustrations of nature, plantations, cities, and harbors as a host of empty, far-off places, ending with a poem praising Earth for its size, diversity, and hospitality toward human beings.[86] The primer, on the other hand, tries to step beyond the level of observation in order to offer insights into

204   GEOGRAPHICAL PRIMER.

James Russell Lowell.

The **yellowish-brown** people belong to Japan, China, and the rest of eastern Asia. They are called the *yellow*, or *Mongolian*, race.

The **blackish-brown** people live on the islands of the Pacific. They are called the *brown* race, or *Malays*.

The **reddish-brown**, or copper-colored people, are the Indians of America. They are called the *red* race.

The **black races** belong to Africa and Australia. The first negroes of our country came from Africa.

EXERCISE. — You have seen persons belonging to the white race. What other races have you seen? Think of one person of each race that you have seen, and write all about him. Tell whether he is a large or a small person, the color of his skin and eyes, what sort of hair he has, anything singular about the shape of his head and features, and what sort of dress he wears.

XIV. — CIVILIZED MEN.
(PART I., PAGES 60-63.)

In our country, in Europe, and in other parts of the world, are states, kingdoms, and empires, with rich cities and educated people who busy themselves about many different occupations. These are *civilized* nations.

In Africa, and some other countries, are men who get all their food and clothing from the wild plants and animals, and know how to build only the rudest huts for shelter. These are *savages*. In still other countries the people till

FIGURE 1.11
"Races of Men" Writing Task, Mary Howe Smith Pratt, *The Guyot Geographical Reader and Primer: A Series of Journeys Round the World*, 1898 (courtesy Special Collections, Monroe C. Gutman Library, Harvard Graduate School of Education).

geography through systematic lessons accompanied by maps and facts. Maintaining a distancing tone throughout, it identifies various "races" that live in the United States with detailed yet categorical descriptions of how they looked: "white men and negroes, reddish-brown Indians, and a few yellowish-brown Chinese and Japanese."[87] This description is followed by a list of six races, as identified in the pseudoscientific human taxonomy that had become common in the West since at least the seventeenth century, and then a writing task for students to complete based on their own personal experiences and encounters (figure 1.11).

Assuming that *The Guyot*'s readers were all white, the task asked them to reflect on their chance meeting with the nonwhite Other and to remark on how different they looked from white people. Though the assignment invited children to think of the individual persons of other races whom they had seen in real life, the exercise mainly asked the students to compile stereotypes and generalizations from the then common theories of race in order to absorb and affirm the content offered by the primer in a uniquely personal manner. Certainly not a creative writing exercise, the task left little room for the inclusion of individual exceptions to the largely fixed racial categories set forth by the primer, let alone any possibility to discard or question them altogether. *The Guyot* expected that children would have internalized the long-established racial order of the world as neutral and natural, including even the common discourse of polygenism that some white American slaveholders of the antebellum years had adapted in defense of slavery. An offshoot of the theory of evolution, polygenism maintains that, upon creation, there were several human races, each with different evolutionary paths and different origins from the others, thus justifying theories that presented the racial inferiority of some races vis-à-vis others as irreversibly divine. As a result of its religious undertones and equipped with a new scientific language since the eighteenth and especially the nineteenth centuries, polygenism came to be widely accepted and wildly invoked in defense of class hierarchies, slavery, and other forms of race-based discrimination in countries such as France, Germany, and the United States, especially in the South.[88]

While continuing to incorporate the human element in geography textbooks, professional geographers had also begun to focus on trade as the ultimate bond between the United States and distant lands and their peoples in the "Commercial Geography" sections of the primers and readers they composed. As the following examples showcase, these writers tended to discard the formerly prominent tone that had passed judgment on human "races" based on their religions and their appearances and according to an assumed divine order. They had become intensely aware that the expanding relations of the United States to various corners of the non-American world hinged on new ways of understanding that world and its inhabitants as distant, potential trade partners of U.S. citizens in a closely knit web of commerce that spanned the surface of the globe. The multivolume *Picturesque Geographical Reader* is a case in point.[89] The series' first edition was published in six volumes. Not surprisingly, with three volumes out of six dedicated to "The Land We Live In," this reader did not differ from *Harper's* in its primary focus on the United States—that is, the closest home in "home geography" and the centerpiece of what mattered most to their young American audiences. The series was written in narrative form and was meant to teach geography to those children who had an active spatial imagination, inviting them to see the United States as in and of a larger planetary framework.

The series' first volume, *At Home and at School* (1890), did not begin with geographical facts about the planet or with the science of mapmaking. Rather, it opened with a practical take on geography in the context of American home—a pedagogic element much favored by educators and youth psychoanalysts such as G. Stanley Hall.[90] Following a tradition already established before the Civil War in juvenile travel fiction by such volumes as the massively popular Rollo series by Jacob Abbott, the reader's first lesson introduced a "trade game" in and around a pond near where the fictional Fred Cartmell and his family lived.[91] Upon Mr. Cartmell's suggestion, the lead character, Fred; his sisters Florence and Nellie; and his father, Mr. Cartmell, were to trade seemingly random goods—such as cotton, wheat, tea, silk, and coffee—between the cotton-rich United States and "a country on the other side [of the pond] where they have no cotton."[92] This act of pseudo-transoceanic trade-adventure merits deeper analysis. First, quickly after the game starts, we learn that the game is playable on two levels of spatial awareness. One is a sense that the so-called closing of the western frontier had placed an emphasis on the "natural" borders of the continental country to the east and the west and turned these borders—that is, the Pacific and the Atlantic as the physical edges of its national geography—into mesmerizing, bustling sites of commerce, points of departure, and ports of arrival. Tightly paired with this is the possibility that Fred's boat in the pond—as a stand-in for the cross-continental ship routes that moved shiploads of U.S.-made trade goods across the oceans—provides: launched into the pond, Fred's boat rapidly rolls up the navigable distance between the cotton-rich United States and the unidentified colonial silk lands, tea lands, and coffee lands that resided in all directions at the end of imaginary sea routes, at once celebrating borders and putting even more emphasis on what those borders signified, safeguarded, and made possible. What is more, the nature of the goods being traded for cotton and wheat—that is, silk, tea, and coffee—suggests the centrality of trade between the United States and almost all continents on earth, especially the then colonies of European empires such as British India or Haiti. Such transoceanic trade, therefore, could have been transatlantic, transpacific, or with the colonies in the Caribbean.

The introductory lesson further reported that at the end of the game the children "hoped, when they grew up, to go across the ocean and learn how people live in other lands."[93] While the Cartmell children stood for all American children (or at least the readers of *The Picturesque Geographical Reader*) as the future of the nation, the lesson also suggested that the history of the nation was preceded by and performed in terms of geographical study, and even that by the end of the century, geography was rendered intelligible primarily in terms of commerce, human relations, and international commercial bonds. In effect, the lesson taught its young readers that in order to see the world and the various groups of people who inhabited it, they needed to graft personal life stories and national histories onto home and world geographies. Just as children needed to grow up before they could go far and meet other people in their diverse dwellings, the lesson suggested, nations needed to come of age before they gained the ability to extend their commercial and diplomatic arms to the world, to enter competitive networks of commerce and diplomacy, or to partake in complex colonizing or civilizing missions.[94]

When the young republic first stepped into the world market at the beginning of the nineteenth century, it was a minor economic entity, mainly importing products from outside its borders. By the time the Cartmells engaged in their trade game, the United States had become the wealthiest nation in the world, and its early-century crude parochialism had been replaced by a tariff-checkered "open door" policy.[95] With a far greater share of world trade, and thanks to increased foreign investment in the country's railroads, manufacturing, real estate, and mining sector, it was soon of paramount importance to white middle-class Americans that their children internalized and learned to maintain the commercial practicalities of geography in tune with the outward reach of U.S. capital to international markets. While the U.S. population had grown from 0.5 percent of world population in 1800 to almost 4.8 percent by 1900, its share of world exports had jumped from 3.2 percent to 15 percent.[96] Such economic growth was meant to continue at a soaring pace, thanks to a generation brought up to fulfill the role encouraged by Fred Cartmell—that of citizens and rulers of the real world.

At Home and at School's short first lesson included a map of the pond that Fred had supposedly drawn in the evening after their adventure (figure 1.12). As the book confided in its readers, this was Fred's first attempt at cartography as a basic yet reliable means of geographical appreciation and recoding of the world. While this exercise began at Cartmell's home, it was a recurring feature throughout the series. In this sense, American children's territorial and racial impressions of the world were governed by an anterior, redraftable cognitive map that they were supposed to have sketched and were going to rely on during the trips they wished to make "when they grew up." What is remarkable is that—as a map of the sea route, *not* the land—Fred's "sketchy map" highlighted the value of sea routes as fixed yet navigable distances

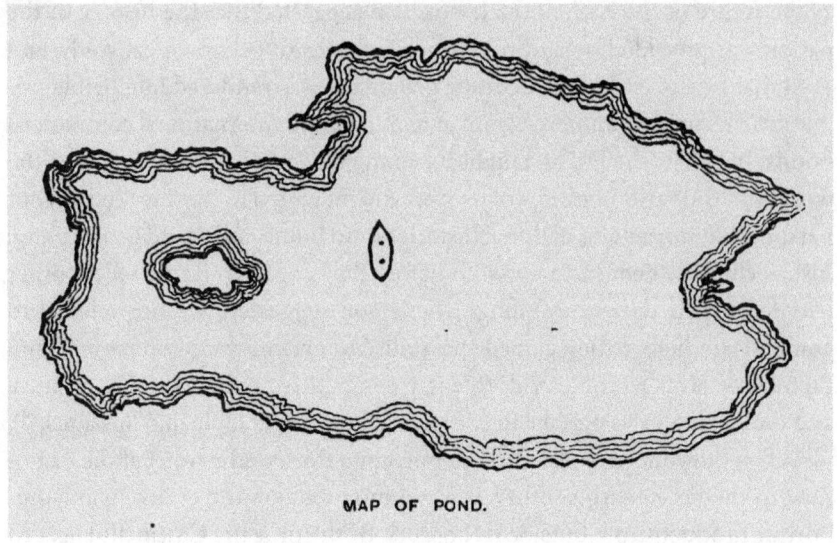

FIGURE 1.12 "Map of Pond," Charles F. King, *The Picturesque Geographical Reader*, 1890 (courtesy of Library Special Collections, Charles E. Young Research Library, UCLA).

between the United States and the rest of the world: the minimum desirable distance between places, polities, and peoples.[97] Thus, this and similar end-of-the-century geography readers coded distance positively, meaning that they prioritized and celebrated it as a means to contextualize the connections between different lands and yet ensure their separation. "At some very basic logic," Edward Said asserts, "imperialism means thinking about, settling on, controlling land that you do not possess, that is distant, that is lived on and owned by others."[98] In Said's reading, traditionally parochial powers such as the nineteenth-century United States could not emerge as global players unless they began to seek, reconceptualize, and shorten distances. In this sense, the continental American empire could not expand its reach to the larger world and achieve global imperial status unless Americans recognized the significance of their geographical distance from the rest of the world and approached it as both doors to keep open and walls to watch over.

In the same fashion, *The Picturesque Geographical Reader* advised American children right at the beginning of their geographical studies to practice drawing sketchy maps of spaces that would matter to them throughout the years leading up to adulthood or to trace routes that connected such spaces to one another and to the United States. During the age of empire, as Said and Benedict Anderson concur, colonial projects were preceded by maps.[99] "A map," asserts Anderson, "anticipated spatial reality, not vice versa."[100] In other words, this map, being a product of the colonial times, "came to be a model for, rather than a model of, what it purported to represent."[101] Likewise, the way Fred's map stood as a model *for* distance, at the same time that it highlighted the commercial interdependence of the nation and the world, reflects Americans' dual approach to the nation's reaching out to the world while honoring its older isolationist tendencies. Either way, before learning geography as a science, children were invited to play the outdoor trade game at a nearby pond in order to learn to imagine and draw maps of larger spaces, conceiving of oceans as navigable routes between the United States, with its powerful naval force, and the world as one huge market.

Since the Enlightenment, illustrations and maps were drawn for, by, and depicting men of fame, with the occasional women who were present on the scene clearly marked as domestic and irrelevant to the cause of geography. This message was stressed in *The Picturesque Geographical Reader*, which made clear that if children wished to fulfill the traditionally masculine responsibility to travel the world, they needed to practice drawing imaginary maps of the world and to take stock of the distances it entailed. Out of step with how home geography conflated the national space with the domestic sphere, by leaving Mrs. Cartmell—the subdued guardian of that most sacred and central of American spaces, the home—out of the trade game (and the lesson), the reader reinscribed geography as an invariably masculine practice.[102] Nevertheless, it should not be overlooked that by the late nineteenth century, American geography benefited immensely from the dedication of quite a few female geographers, who developed geographic theories and compiled successful school primers and expert geography books. In fact, at a time when medicine and astronomy were still quite closed to American women, geography, at least on the conceptual level, had opened its doors to women. Sarah S. Cornell and Ellen Churchill Semple were members of a prominent generation of professional geographers whose works were read, critiqued, and taught at different venues across the nation.[103]

Overall, the game offered a simplified, anthropocentric, matter-of-fact definition of geography, primarily as the study of the planet, its fragmented distant landmasses, and its people, much as *Harper's* did: over-the-pond people and their lives scripted as out of the ordinary and different from those of Americans; their economic needs as the engine that sustained international markets and bound them with commercial ties to the United States; and their relative distance from the United States as substantial but traversable. After all, young Fred's boat could and *did* navigate this geographically naturalized and commercially significant distance in a considerably short amount of time. In essence, the reader confirmed and qualified one foundational definition of geography as "a way of distinguishing 'here' from 'there,' without which little sense can be made of human experience."[104]

IN 1898, AND IN THE CONTEXT OF NEW, outward-looking expansionist impulses across the political board, W. E. B. Du Bois invoked a metaphor in reference to the incontestable interdependence of the United States and the rest of the world—and, going even a step further, to their mutual indispensability: "On our breakfast table lies each morning the toil of Europe, Asia, and Africa, and the isles of the sea; we sow and spin for unseen millions, and countless myriads weave and plant for us; we have made the earth smaller and life broader by annihilating distance, magnifying the human voice and the stars, binding nation to nation, until today, for the first time in history, there is one standard of human culture as well in New York as in London, in Cape Town as in Paris, in Bombay as in Berlin."[105] Moving attention away from the insuperable differences between the types of labor that sustained the global web of colonization mapped out here, Du Bois placed emphasis in this passage on the imagined reciprocity between the colonizer and the colonized; at the same time that the colonizers—in fact, their poorly paid working classes— "sow and spin," he maintained, the colonized "weave and plant." Referring to the nodes of the global colonial world and the most American of cities in terms of spirit and conduct in the same breath, Du Bois's statement commemorated a decisive moment in U.S. history: the Spanish-Cuban-Philippine-American War. In less than a hundred words, Du Bois juxtaposed the map of the emerging overseas empire with that of European empires, pointing at once to the changing semantics and the enlarging context of everyday life in the United States in the same manner in which Dodge and Kirchwey pointed in geography classes to the American "breakfast table" as the departure point from which American children were to see "the whole world...in its relation to the[ir] communit[ies]."[106] This was a new draft of geography, bent to fortify the old but also to forge new bonds between didactics, cartography, and empire.

But what forms did this new geography take as it molded children's views of the world outside the schoolroom?

CHAPTER TWO

Quiet as Mice
Dissected Maps, Domestic Fun, and the World in Pieces

In 1889, Elizabeth Jane Cochran (1864–1922), an investigative journalist and reporter for the *New York World* who wrote under the pseudonym Nellie Bly, set out for a trip around the world. Inspired by Jules Verne's *Around the World in Eighty Days*, Cochran made plans for her trip in 1888 and set off the following November.[1] Beating Jules Verne's fictional record, it took Cochran only slightly longer than seventy-two days to complete her trip. The news of her trip was overwhelming. In effect, as Joyce E. Chaplin maintains, Nellie Bly's trip had proved so lucrative and stimulating that several other Americans set out on similar trips around the world.[2] These included, among others, seaman Captain Joshua Slocum, who sailed around the world from 1895 to 1898, and Latvian immigrant Annie Cohen Kopchovsky, who circumnavigated the globe by bicycle.[3] The public excitement over Nellie Bly's 1889–90 trip around the world had roots in an ever more sophisticated consumer culture, which had begun in the last quarter of the nineteenth century and which capitalized on play as a profitable market niche and a wholesome family pastime.[4] In 1890 alone, McLoughlin Brothers made more than one version of the board game Round the World with Nellie Bly, which took an excited American public on a virtual world tour. To be played by two to four players, the game enabled Americans to turn into "globe circlers" and to break the speed record for crossing the Atlantic. Available for purchase through mail-order catalogs in urban and rural America, the game made it possible for a large number of Americans to imagine themselves touring the world and breaking the record of the fictional British globe-trotter Phileas Fogg—not only as individuals and families but, ultimately, as a nation.[5] Round the World with Nellie Bly was a variant of older board games such as the moralistic Mansion of Happiness (first copyrighted in 1843 by W. & S. B. Ives) and Innocence Abroad (copyrighted in 1888 by Parker Brothers), which capitalized on the original popularity of Mark Twain's *The Innocents Abroad, or The New Pilgrims' Progress*.[6] In the aftermath of Nellie Bly's record-breaking trip around the world, Americans hoped that the journey and the publicity attending it could, rather literally, put them on the map—and they did. In McLoughlin catalogs, board games such as *Trip Round the World; A Dash for the North Pole; Across the Sea, or A Trip to Europe;* and *War at Sea, or "Don't Give Up the Ship"* were alternatively referred to as "map games." Marketed to children with the express purpose of teaching them geography or history "without conscious study," these games treated maps as game boards where education met entertainment.[7]

At the same time, as new waves of migrants from southern and eastern Europe continued to reach the shores of the United States, and unfettered urbanization raised serious moral and hygiene issues, which some better-off white Americans attributed to working-class and migrant families, an overwhelming number of American households began to wall themselves off from the larger society. Many homes turned into indoor sites of recreation, with increased emphasis on reading together, performing home theatricals, and playing parlor games.[8] Urban children were kept indoors and away from the supposed vices of the street. This became a trend that gave rise to the need for entertainment for the longer evenings, which children and adolescents now spent at home. As play came to be more closely associated with a robust experience of childhood rather than sinfulness and idleness, competition among major East Coast toy manufacturers and publishers such as Milton Bradley, Parker Brothers, McLoughlin Brothers, and Selchow and Righter contributed to the proliferation of new games and methods of manufacture, and increased attention to social and political events as potential topics for new games. Having invested in children as consumers for nearly a century, these companies helped invent the modern artifact of childhood. As the century waned, they were producing a massive body of both pedagogical and non-pedagogical playthings for children as well as for nostalgic adults. As one example, McLoughlin Brothers, which had begun by printing religious tracts, valentines, and comic almanacs in the late 1820s, had by 1850 grown exponentially and was, by the century's end, an expert in child entertainment. In the second half of the century, McLoughlin Brothers printed elaborately illustrated original or pirated chromolithographed toy books, paper dolls, ABC books, alphabet cards, and card and board games, which they marketed to American families—that is, to be bought by parents and consumed, mainly but not exclusively, by children.[9]

In such relatively fun-oriented leisure times, domestic diversions including not only geographic board games but also dissected maps, picture puzzles, and geographical riddles confounded the otherwise mathematical, one-to-one representational relationship between cartography and the world. In getting close to the playing child outside the confines of the schoolroom, the present chapter examines the ways that dissected maps (cartographic jigsaw puzzles consisting of maps mounted onto cardboard or thin wood and cut into pieces) and, more generally, geography-themed picture puzzles ludified the relationship between maps and the spaces they represented. As I will argue, depending on their producers' commercial interests, political worldviews, and status in the consumer market, such visual media enjoyed varying levels of authority and authenticity. In the ensuing discussions on engraving, dissecting, and playing with world maps, and following a brief survey of the origins of geographical games in imperial Europe, I examine dissected maps as objects that aimed at teaching lessons in world geography through home-based entertainment. In the following section on playing with dissected maps, I explore the range of meanings and uses of geographically scripted material in each of the three senses that Robin Bernstein invokes: the documented, the probable, and the possible.[10] My aim is to examine these "texts" beyond the intended original uses that were determined by manufacturers and educators by considering the mostly underexamined range of probable and possible uses these playthings and educational objects were put to by their young owners.

European Dissected Maps: A Brief History

Also known as "puzzle maps" in the second half of the nineteenth century, dissected maps were marketed as educational toys on both sides of the Atlantic and were for the first century and a half of their existence expressly directed at children.[11] Although there is no consensus as to where and when jigsaw puzzles were made for the first time, puzzle historians agree that as the first modern-day puzzles, dissected maps were originally manufactured by Madame Jeanne-Marie Leprince de Beaumont, who sold *cartes de géographie en bois* at one half guinea to the young girls who attended her school on Henrietta Street, Cavendish Square, London, as early as 1759.[12] Jill Shefrin's careful examination of Lady Charlotte Finch's life and career as the governess to King George III's children reveals that Madame Beaumont, a London-based author and educator, was probably the first producer of maps that were dissected for educational purposes.[13] Further research by Shefrin suggests that dissected maps were in a matter of years masterfully perfected, inventively diversified, and successfully marketed to children of the British royalty by John Spilsbury (1739–69), apprentice to King George III's geographer Thomas Jefferys (figure 2.1). As Linda Hannas reports, Spilsbury's second trade card advertised "All sorts of Dissected Maps for Teaching Geography" and listed twenty-nine dissected maps of Britain's colonial possessions around the globe,

FIGURE 2.1 *Europe Divided into Its Kingdoms,* John Spilsbury, 1766–1777 (© The British Library Board, Cartographic Items Maps 188.v.12).

including Jamaica, the "United Provinces" (now Uttar Pradesh in northern India), and the "West India Islands."[14] Other than maps of the British Isles themselves, maps of Africa, the Americas, and Jamaica—in general, maps of those parts of the world that mattered most to the British Empire—were among Spilsbury's earliest dissected maps. Madame Beaumont's invention and Spilsbury's ensuing success in the market were met with great fascination among the British upper classes, so much so that Spilsbury was commissioned by Lady Charlotte Finch to manufacture dissected maps for the King's thirteen children to study geography through play.[15]

Spilsbury's master, Thomas Jefferys, was appointed royal geographer to the Prince of Wales, and later King George III, from 1746 until his death in

1771. During this period, Jefferys engraved and printed several maps and atlases for King George III and for other official bodies in the British government. Among his major achievements in this honorary position was the engraving of several maps of various parts of the North American continent, including *The American Atlas*, which included *A Map of the Most Inhabited Part of Virginia* and *A Map of the Most Inhabited Parts of New England*. Jefferys's *American Atlas* was reprinted and sold mainly to a war-frenzied British public immediately before, during, and even after the Seven Years' War (1756–63), which was fought between France and Britain on three continents. As an apprentice to Jefferys, Spilsbury had access to all the resources that his master's position as royal geographer offered.[16]

Despite Spilsbury's death at the age of thirty, dissected maps stood the test of time. Over the following decades, British and later German, Dutch, and French manufacturers scurried to diversify the themes, including both didactics and religion.[17] Cubic and metamorphosis puzzles proliferated in the first quarter of the nineteenth century, adding diversity and complexity to the previously flat educational dissected maps.[18] What is more, new cutting and dyeing techniques, the use of cardboard instead of hardwood to mount the maps, and the use of lithography instead of engraving to produce the images that were developed in these European workshops made the manufacturing process easier, faster, and cheaper than in Spilsbury's time.[19]

Whether they were Spilsbury's invention or Madame Beaumont's brainchild, by the beginning of the nineteenth century, dissected maps were being imported to the United States from toy workshops in London, Paris, and Nuremberg—workshops that mostly ran on slave and servant labor.[20] Once American manufacturers started producing dissected maps, they mainly followed the British model and aimed to manufacture dissected maps at lower prices than their European counterparts by applying innovative techniques, using cheaper materials, and diversifying the themes and images they chose. At their peak of popularity in the United States and especially in the postbellum era, dissected maps were manufactured by a great number of toy manufacturers in major East Coast cities, firms such as Milton Bradley of Springfield, Massachusetts, and McLoughlin Brothers of New York. Consequently, U.S.-manufactured dissected maps and picture puzzles were available to children from a wide range of backgrounds, serving their interest in geography, history, natural history, and related themes. For example, McLoughlin Brothers (1828–1920)—the most prominent nineteenth-century American children's toy manufacturer and book publisher—managed to manufacture lower-priced dissected maps by following the European technique of mounting them on cardboard. Furthermore, for quite a long time, McLoughlin Brothers and other toy companies "cheerfully sliced [maps] up with no regard to the boundaries"[21]—a considerably less expensive mode of "dissecting" than cutting along the irregular shapes of political borders.[22]

American Dissected Maps: Home Geography and Reinscriptive Cartography

Thanks to the immense diversity of the themes they offered, picture puzzles had become increasingly popular by the last quarter of the nineteenth century. By this time, they were no longer unidimensional tools of education. Rather, U.S.-manufactured dissected maps and picture puzzles reflected toy makers' fascination with the American lifestyle and the ways it interacted with nature, especially on the ever-eroding western frontier. Thematic diversity in dissected maps and picture puzzles was so widespread that as late as the 1890s, and long after dissected maps had entered American children's lives as a common and popular toy, McLoughlin Brothers still categorized their dissected maps and picture puzzles in their confidential price list books under "novelties."[23] However, the rise of image-based jigsaw puzzles did not dampen dissected maps' popularity among teachers, parents, and children at play. On the contrary, many American puzzle makers continued to market dissected maps as educational toys and remained faithful to the basics of geography and cartographic precision. In effect, dissected maps had proven so useful as educational materials that by the 1870s, the Texas Institution for the Blind had begun to acquire dissected maps of Europe and the United States with "raised outlines"—a technology that had made them suitable to be used by visually impaired children who could not otherwise make use of cartographic maps.[24]

Whether used to keep children silent as they played indoors in the evening or to teach them world geography or natural history, toys are hardly ever free from intertextuality. This means that their use at any given moment intertwines with their earlier forms and editions, the play practices associated with them, and their parallel uses in other contexts.[25] Conceived in such a web of intertextuality, toys can be transplanted into new contexts where they gain new meanings once children of different socioeconomic backgrounds, genders, and races play with them—a process Jackie Marsh and Elaine Millard refer to as "metanarrativization." In the case of dissected maps, metanarrativization occurred most prominently because these objects were maps (tools in the adult world) first and toys (playthings in the world of children) second. As the box cover of the 1880 edition of the *Dissected Outline Map of the United States of America* by Milton Bradley confirms, dissected maps were meant to form "a complete object lesson of the geography and natural history of the United States."[26] A child playing with such a dissected map would then necessarily engage with the hidden codes with which adults design toys. This included the design principle of simply decontextualizing and repurposing an adult's tool as a children's toy. Subsequent to this repurposing, a second side to metanarrativization would emerge: as children learn about their country's natural history or practice lessons in geopolitics, a reversal in the tool-to-toy process materializes that reverts those very toys back into tools of a different order. Adapted as educational tools, dissected maps were more than toys. Indeed, they were tools that socialized children into their future roles as the next generation of geographically privileged adults in charge of a global empire.

Cartography as a process entails the cartographer's need to first imagine the spatial whole that they wish to draw, as well as its constituent parts, and to then scale those elements down not only according to cartographic conventions and mathematical rules but also with an eye to the given map's purpose and audience before a map is finally engraved. That same process is true for children playing with dissected maps: children needed to first think of the whole space that their geographical toy represented and to then come up with that whole by patiently putting together the pieces which the toy makers had cut apart. By piecing together the whole out of the cut-down pieces, even when a small piece or two went missing forever, children would engage in reproducing a larger representational whole out of irregularly shaped pieces every time they played with a puzzle. Such an entertaining practice—which I term an act of "reinscriptive cartography"—could underscore the cohesion that cartographic representations such as maps had always meant to offer: the whole as the cognitively comprehensible geographic ideal behind any act of cartography. As Anne D. Williams posits, "To bring order and beauty out of the chaos of hundreds of disconnected pieces" has always been the motivating drive behind solving jigsaw puzzles, no matter how complicated and time-consuming completing one might be.[27]

As educational playthings that were meant to reinscribe cartography, dissected maps enjoyed distinctive features. To begin with, they were popular not in the classroom but in the domestic space—keeping children "quiet as mice" (figure 2.2). Aware of their potential to enable pedagogy but also to permit play (including destruction and reconstruction of the toy), American puzzle manufacturers advertised dissected maps as both amusing and pacifying.

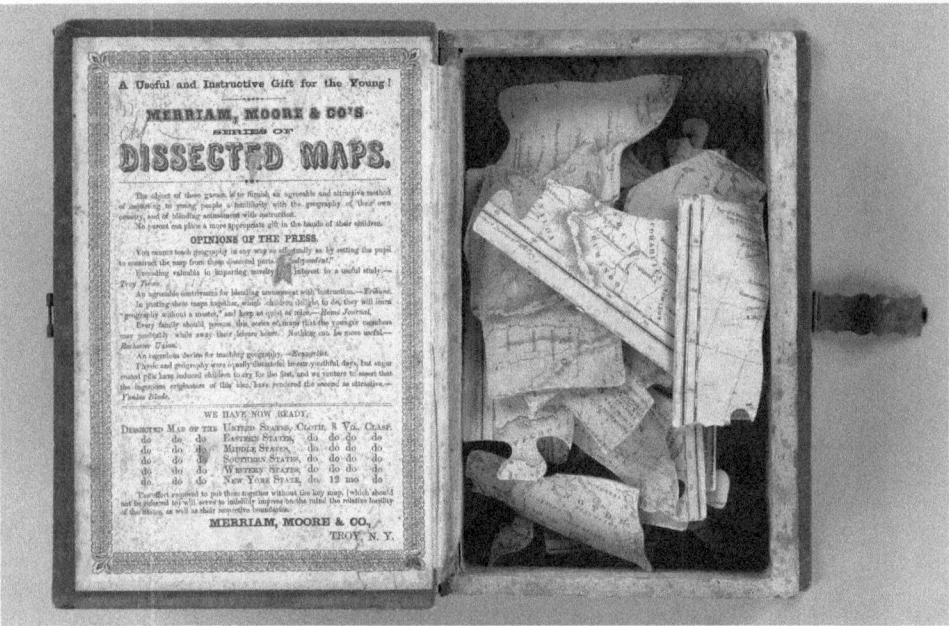

FIGURE 2.2 Inside cover of the box of the *Series of Dissected Maps*, Merriam, Moore & Co., 1853 (courtesy of David Rumsey Map Collection).

These factors mattered especially to urban, middle-class parents, who by the end of the nineteenth century were busy outside the home and had little time and energy to spend with their children in the evenings. With the advent of urbanism as the norm rather than the exception, affordable and amusing board games and dissected puzzles further appealed to well-to-do American parents who focused on regulating their children's bodies and pastimes by amusing them in the safety of their homes rather than letting them play in city streets.[28]

Regardless of the form it takes and the material objects it involves, child play is certain to have short- as well as long-term influences on children's imagination and perception of themselves and the world and on how children express their fantasy-driven relationship with the world around them or the worlds beyond their reach. Moreover, as Brian Sutton-Smith maintains, the history of the past three centuries suggests an intense "domestication" and regulation of children's play by adults who hope to regiment children's playtime, corporeal experiences, and imagination. Such domestication, Sutton-Smith asserts, results from the commonly held belief by generations of adults that children are the society's future and, as such, need to be molded into the best shape possible through education and entertainment.[29]

As a result of the playful relationship that some children developed with dissected maps first and foremost as playthings, these toys joined efforts with geography primers, mass-produced maps and globes, and home atlases to popularize geography as a school subject. Dissected maps allowed young players to spend time practicing what they were supposed to have learned in their geography lessons. In other words, dissected maps were meant to turn learning an increasingly important yet not always engaging school subject into fun, or at least to add an entertaining edge to it. While playing with dissected maps, children would grapple intellectually and physically with the puzzles' irregularly hand-sawed or guillotined pieces. On the physical level, this meant that children touched the small pieces of the map and held them in their hands (even that they would momentarily dip the pieces' tips in their mouths while thinking and searching), paying more attention to the pieces' shapes, edges, and borders than to what was visible on each one of them. On a cognitive level, this involved maneuvering their minds' eyes over already internalized maps of the places that they pieced together on the floor. Children could practice scaling up, or down, the maps they had in their minds in order to put the pieces together following the normalized order the original maps dictated. Once maps lay open in front of them, from either their geography books or home atlases, children could match their scales to those that were used in producing the dissected maps in the first place and piece them back into meaningful spatial unities—that is, into cartographic maps—before then intentionally destroying them to start to play again.

In his grid of spatial practices, David Harvey categorizes cartographic maps under "representation"—visual "perceptions" that capture expansive spaces on varying scales and make them visually and cognitively accessible to the human mind (see figure 1.1).[30] Following suit, dissected maps seem to be no different than cut-down multiplied maps that represent the same places that undissected original maps do. Still, similar to colonial projects of territorial conquest—whereby colonizers would select a distant or nearby land, annex it, and cut the surface of the earth with arbitrary borders between nations and real and imaginary routes connecting colonies to one another *and* to the metropole—the act of selecting an existing map of a specific region and of dissecting it into an educational plaything is an example of a spatial practice of a more subtly manipulative nature, which Harvey calls "domination and control of space." Once what Harvey calls "territorial imperatives" enter the equation in dissecting maps and playing with them, mastery over cut-up spaces begins to overrule geographical accuracy and scalar precision. In this sense, then, dissected maps are both the ideological offspring and the material agents of colonial violence.[31]

As an activity in which children could practice both tactile and cognitive "domination and control of space," piecing together unitary wholes out of a fixed number of irregularly shaped pieces has spatial significance on at least three key levels. First, other than keeping them silently occupied without the need for adult supervision, the ultimate goal in encouraging children to play with dissected maps was to remind young people of the importance of the individual pieces in making the whole possible. These geographical playthings taught them that without the pieces, no spatial whole was conceivable. This mode of play would empower pieces as the building blocks of the whole in the child player's spatial imagination of the world and in this way align the child's habits of spatial thinking with the world order that the United States desired geopolitically.

It would appear at first glance as if dissected maps were no different from maps that were cut into smaller pieces. In their two-dimensionality, they too distorted scales and excluded or exaggerated topographical features, exercising immense degrees of power over what map readers had, or did not have, access to. On closer inspection, however, it becomes clear that each piece of a dissected map is itself the map of a smaller region. This quality is especially important because by the final decade of the nineteenth century, dissection took place mainly along political borders, assigning to the pieces an immense degree of geopolitical weight in the American geographical imagination. Put differently, dissected maps differed from their cartographic siblings both in their materiality and the message they conveyed. On the one hand, dissected maps complicated the way that children "read" maps by looking at, touching, and playing with them as a toy whose charm emanates from a tangle of redoability, fragility, and multitudinousness. On the other hand, dissected maps could broaden children's focus on the cartographic whole—whether the national or the global—by primarily drawing their attention to the significance of the *parts*—be it the regional or the local—that make that whole possible. In this manner, the intended use of dissected maps closely resembled the ways home geography as a pedagogic approach to world geography wove the accessible study of local/home geographies into a larger, eclectically quilted global network of other localities and homes. In other words, the issues raised by playing with dissected maps as geographical toys are also object lessons about restoring a predominantly colonial world order that was made sense of in the 1890s and beyond as "home geography," where each home had to be placed in its exact place in a seamless, uniform global fabric and yet at a distance deemed appropriate from other homes.

Second, depending on which region, country, or continent the dissected maps represented—be it the whole world, the United States, or the state of New York—in getting to know images of those parts of the world through play, American children would further learn about the normative geographic urgencies of the empire. As David Brody suggests, "Visuality played a key role in establishing empire."[32] As a great number of surviving puzzles from the late nineteenth century suggest, the choice of maps selected for dissection clearly reflected the changing national priorities of the United States. As Schulten maintains, "How comprehensively a region is mapped indicates what is known of the region and also its perceived importance."[33] As maps' imperial siblings, dissected maps proved to be no exception to this rule. Placing emphasis on visuality and materiality of imperial projects, dissected maps labored to communicate common perceptions of geographical import from adult Americans to the children who were the intended beneficiaries of the regions they displayed. This was especially true during the wave of overseas territorial acquisitions that followed on and exceeded the scale of westward expansion. Following U.S. expansionist impulses outside the North American continent in the wake of the Spanish-Cuban-Philippine-American War, carefully selected pieces of the world map gained unprecedented significance rather immediately and were suddenly considered unitary wholes worthy of zooming in on, further scaling down, and dissecting. As a consequence of the controversial involvement of the United States in the Philippines since 1898, for instance, and as maps of the archipelago proliferated on the U.S. market, that specific part of the world became important enough for Americans to choose as an individual map to dissect into a puzzle (figure 2.3).

Finally, while maps in school primers and home atlases were meant to visually acquaint children with the world, or with any units smaller than that, dissected maps focused on children mastering the world, both visually and in a tactile manner, by presenting them with the chance to hold its smaller parts in their hands and to play with them kinetically. Based on the material nature of geography, the future prominence of America as a rising world power in the twentieth century was thus facilitated by encouraging American children to practice mastering its two-dimensional representations by playing with them even after the geography lesson was over. As educational playthings, dissected maps offered little space for children to imagine the world *differently* from the mathematically scaled-down cartographic maps that toy manufacturers chose to dissect. In other words, dissected maps—especially those with interlocking puzzle pieces—left little room, if any, for imagining an alternative

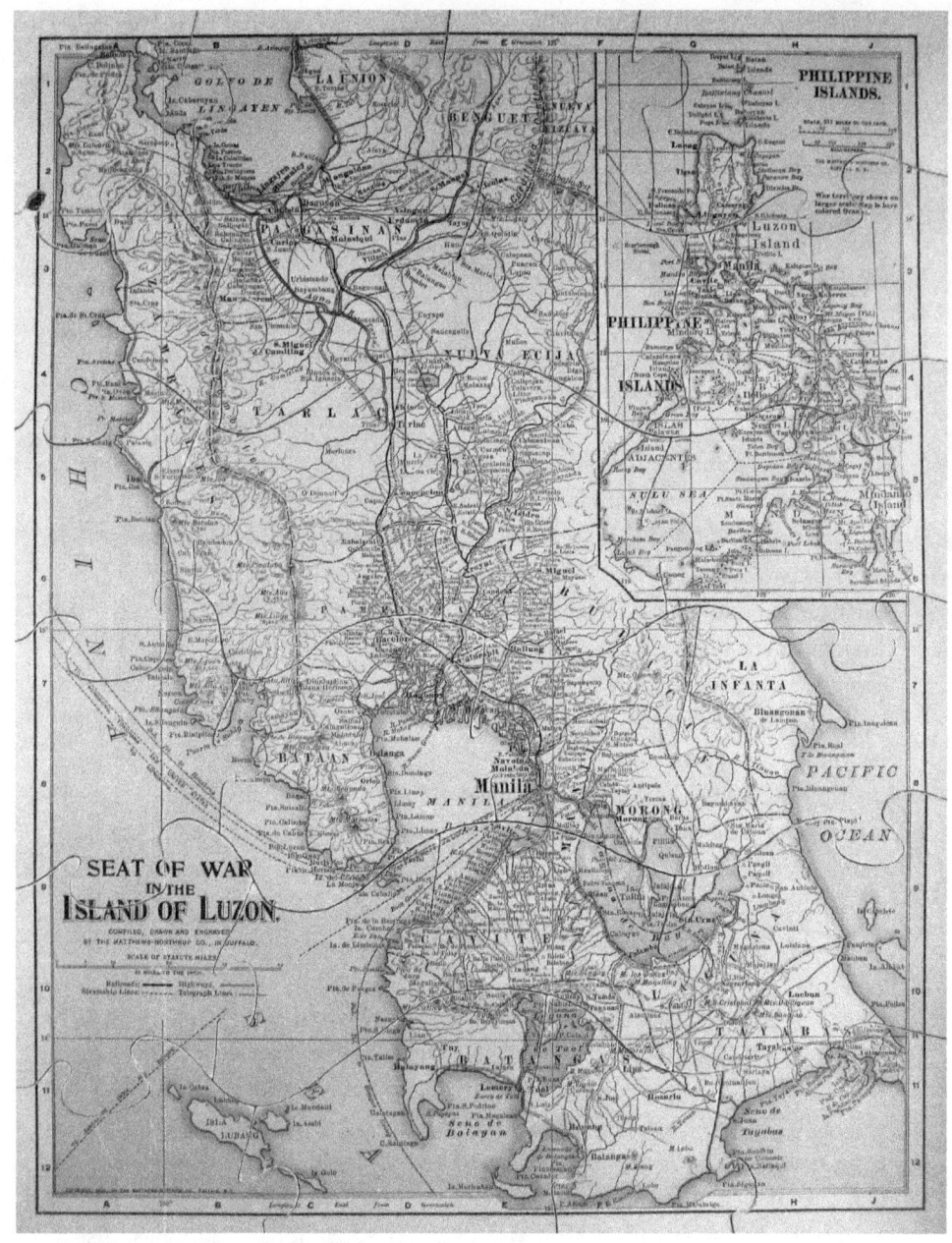

FIGURE 2.3 *The Philippines Dissected Map*, Parker Brothers, 1900 (courtesy of The Liman Collection, New-York Historical Society).

order in solving the puzzle, or for imagining the United States as anything other than a prominent part of the world. Therefore, the outcome of hours of comparing the map—stuck to the inside of the puzzle's box—to the pieces on the floor seems to have been inevitably the same: piecing together a cartographic whole, be it the world or its various parts, as the child's plaything.

For this reason, dissected maps might seem to have limited children's creativity and disallowed them from imagining an alternative world order. However, the world order they *did* present was far from static and allowed for an exceedingly more complex understanding of international relations than the model imposed by formal didactic maps. To be sure, a full atlas of the world in dissected form did not exist in the nineteenth century—a fact that confirms that dissected maps did not have to abide by the comprehensive, global pretensions of the school primer and the home atlas. Inevitably, toy manufacturers' choice of dissection followed the existing pool of maps available on the market. Toy makers turned existing maps into toys, but they generally did not print the maps themselves. Consequently, in choosing from that pool, American adults imposed the nation's geopolitical imaginaries of the world on children's play activities. Ultimately, the very act of picking a certain region to dissect could teach children a subtle lesson in geopolitics, suggesting an alternative metageographical world order by focusing attention, however temporarily, on a region other than the United States.

Similarly, while the use of different colors on dissected maps imposed a host of limitations and restrictions on children's perceptions of the world, it could also open up spaces for associative imagination of different regions on the maps in a manner that sometimes deviated from the geopolitical lessons that the agents of the globally emerging empire had in mind. Color choice on late nineteenth-century maps followed practices that had been formulated by, silently agreed upon, and aggressively marketed as geographic truths by European imperial cartographers and engravers—a hierarchizing visual technique that was applied to both maps and dissected maps first in Europe and later in the United States. Different colors distinguished different states or regions of the nation and different parts of the world, including individual continents and countries. In the case of world maps, different colors were used more or less systematically to bind together Old World empires and their respective colonies and yet to distinguish each imperial family from other imperial families and from alluring unexplored regions, the so-called terra incognita, scattered through Africa and Australia. According to Benedict Anderson, British imperial maps developed color codes for the colonies of different European empires, using pink-red for the British colonies, purple-blue for the

French, and yellow-brown for the Dutch.[34] "Dyed this way," Anderson asserts, "each colony appeared like a detachable piece of a jigsaw puzzle. As this jigsaw effect became the norm, each piece could be wholly detached from its geographic context."[35] As a result of this "jigsaw effect," such a seemingly innocent practice—partly aesthetic codification and partly epistemological typecasting—was carried over to the other side of the Atlantic, where it was adopted by cartographers to single out colonies in their arbitrary geographical places on the map. At the same time, however, color coding made the colonized pieces of the maps relatively movable in relation to the metropole, a practice that, Anderson believes, paved the way in the twentieth century for the formerly colonized to envision decolonization and to ultimately realize it.[36]

Dissecting, Packaging, and Playing

Whether dissected maps appealed as toys or failed as pedagogical tools, and whether they induced creativity or restricted spatial imagination, what is at stake in our understanding of the geopolitical inferences coded into dissected maps is the conversion that adapts the nationalistic/imperialistic attitudes that adults envisioned for the nation's future to the visual/cognitive aptitudes this mode of playful learning exacted from children. Such slippage was meant to result in a heightened sensitivity in children toward the changing dynamics of the global spatial status quo, including the shifts in the role of the U.S. empire in world affairs. In turn, while playing with their dissected maps, children would try to give order to the scattered pieces on the floor and to come up with identical, accurate cartographic images of the entire world, this or that continent, or the United States itself.

At the same time, toward the end of the century, puzzle manufacturers took every opportunity to bring in a larger context in order to more forcefully position the United States as a "nation among nations." Moreover, they consistently managed to teach the significance of strategic absences and presences through illustrations on the box cover or, in the case of two-sided maps, on the other side of a dissected map of the United States itself. *Clemens' Silent Teacher Dissected Map of the United States* (1883–88) is a case in point. Earlier editions of *Silent Teacher* either have plain, unillustrated box covers or show young boys and girls playing with the toy in their lavishly furnished living rooms. The puzzle's first illustrated box shows a domestic scene where four young white children and their pet cat play with the dissected map by the fireside in the company of carefree adults who are engaged in their own pastimes: Two women from different generations sit silently embroidering in

FIGURE 2.4   Original box of *Clemens' Silent Teacher*, Rev. E. J. Clemens, 1883–1888 (courtesy of The Strong, Rochester, New York).

the foreground, while an older male figure, perhaps the grandfather, occupies himself with a newspaper, his back to the scene (figure 2.4).[37]

Echoing the hope expressed in the advertisement for the *Harper's School Geography* discussed in chapter 1—that knowledge of world geography in children would lead to reading newspapers as adults—a parallel presents itself between the role the newspaper plays for the older man and what the dissected map is meant to do for children. While for the better part of the nineteenth century and well into the next, geography was considered a masculine, outdoor adventure, this illustration points to at least two things in keeping with the trend to domesticate children and their play during the late nineteenth and early twentieth centuries. First, it suggests that by entering school curricula on a national level, geography had further entered a new era of unisex appeal and utility. Second, the image of domestic play implies that although nominally a masculine occupation, geography—as it was learned at home and under the supervisory role the older women (mothers of the nation) assumed over children's games—was a vital undertaking that interlaced the familial and the domestic with the public and the political.

In contrast to its 1883–88 edition, the 1893 edition of the *Silent Teacher* bears an elaborate color illustration that underscores the increasing attention the nation paid to the world in which it sought prominence (figure 2.5). The illustration shows an American gentleman (with the face and clothing of a toddler) who is handing a box containing the dissected map of the United States to a gratefully bowing Chinese figure with his right hand while pointing to a map of the United States on the ground with his left index finger. The box-lid illustration sets an all-male scene in a semitropical wilderness in the background and the map of the United States in the foreground. Six adult male Europeans, fully and colorfully garbed in their national costumes, also stand in the foreground, looking in the direction of the American figure with eager eyes. Among these, a British, a French, and a German gentleman stand with boxes of the *Dissected Map of the United States* in their hands. Most conspicuous here are the ways imperial seniority surfaces as a visual element in the details of this illustration: given their familiarity with the former colonies before and right after the American Revolution, the figures representing the comparatively older French and British Empires (as former colonizers of parts of North America and hence familiar with its geography) comfortably hold their boxes under their left arms, while the representative of the younger German colonial empire holds his box deferentially at a short distance in front of him and with both hands.

Stacked next to the American gentleman, the boxes bearing the labels "Spain" (with its own colonial history in the Americas), "Turkey," and "Russia" lie unattended on the ground. Do the Spanish, Turkish, and Russian figures need to stand by until the American figure is done with the geopolitically more important China before they, too, receive their boxes? Even further removed from the center, placed behind these sharply outlined figures and off to one side of the scene, a large number of shadowy figures stand. Perhaps a sign of their diminishing state of "civilization" or lack of "civilization" altogether, all that is visible are their hats dully projecting from a short distance in the background. Among them are representatives of other nations that are supposedly less civilized and less important to the United States, including Middle and Far Eastern nations and "tribes." Unlike the European gentlemen present at the scene, none of these figures is depicted with a cheerful, eager countenance; excluded from the scene except as fading silhouettes, they appear as lackluster, outlying, hard-to-identify bystanders rather than as keen partakers in the scene's central activity, bequeathing the world access to geographic knowledge about the rising U.S. empire. There are no boxes in the hands of these shadowy figures, nor are there any near them on the ground.

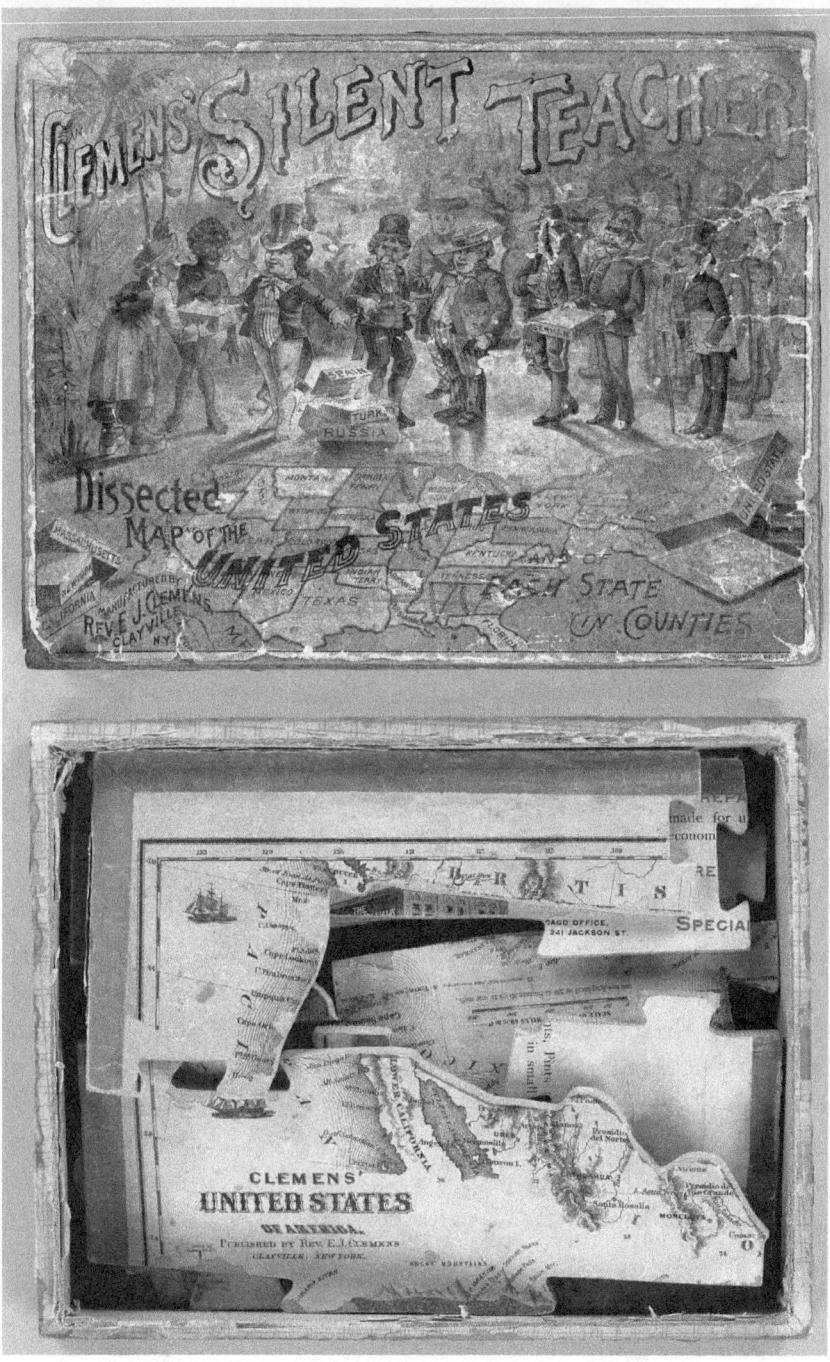

FIGURE 2.5 Box cover of *Clemens' Silent Teacher*, Rev. E. J. Clemens, 1893 (courtesy of David Rumsey Map Collection).

Assumed to be at once illiterate and incognizant of the significance of the highly momentous affairs around them (and consequently judged unworthy of receiving the American token), these figures—nationless and hence map-less, irrelevant to the cause of geography and, by extension, that of geopolitics—recede into the wilderness.

Following common civilizational stages identified by the Dutch Conrad Malte-Brun in 1810—recurrent references to which were made in geography primers of the time on both sides of the Atlantic, including Samuel Augustus Mitchell's *System of Modern Geography*—this illustration is a graphic product of a racialized narrative of hierarchies that allowed the United States to enlist itself as a fully mapped and therefore civilized member of the "family of nations."[38] The scene signifies the supposed openness with which the United States interacted with other parts of the world, not simply by studying their geography, but also by offering them the chance to get to know U.S. geography. By the 1890s, westward expansion had more or less reached its land limits in the West. The imagined geography of the nation, in other words, was more or less stabilized, leaving the United States ready—in this depiction—to join older imperial powers on equal footing as an emerging overseas empire. Certainly, this image was legible to young Americans who had developed a solid sense of world geography affairs through their access to more affordable maps, geography schoolbooks, and home atlases.

As the 1893 edition of the *Silent Teacher* attests, the complex relationship that the United States, as a young yet purportedly enlightened nation, perceived itself as having with other nations started but did not stop at home. At the front of the scene, and giving the image its visual focus, stands an eager and excited figure, sharply colored like other Europeans and yet (unlike almost all the other figures in the scene) half-naked and hatless. This supposedly young African adolescent stands nearby as the even younger American figure hands the box to the Chinese figure. There is no box labeled "Africa" in his hands or anywhere on the ground. This figure might have stood for the African American bystander whose function was presumably to protect his white American "master" with the long spear in his hand. However, the inquisitive look in his eyes and his stereotypically premodern countenance and costume (following common European representational strategies in depicting African people in the nineteenth century and beyond) suggest that he is meant to represent Africa as an undefined and undifferentiated whole—a terra incognita simplified to comprise the African continent in its entirety. Remarkably, the boy's sharp outline and his placement in the foreground of the scene, next to the representatives of masculine, "adult" European empires and China, seems to unsettle the common assumption that nineteenth-century Americans had an unalterably firm belief in civilizational hierarchies. Contrary to that, this image confirms that Americans were willing to embrace Africa as a potential partner when it came to new economic opportunities and the geopolitical exigencies of their expanding empire.

In fact, admitting China into the camp of the "civilized" as its latest member while "savage Africa" eagerly awaits and watches the ritual suggests how geopolitics and the hope of finding economic niches in the "non-civilized" world could, and did, modify white American adults' otherwise firm belief in a litany of insurmountable differences between "races." As the illustration confirms, the United States was now willing to make exceptions in the case of a stand-alone China with its huge market, and in the case of a purportedly welcoming Africa with its supposedly endless natural resources. In other words, by century's end, the United States was at times willing to ignore existing hierarchies and to define its civilizational distance from certain regions according to its fluid geopolitical interests. It was hoped by American expansionists that, upon receiving the American token, China would be understood as a sovereign nation, equal to and independent of European and Japanese oligopoly over its market and politics. Following their far-reaching commercial interests and expansionist policies outside the continental borders of the nation, Americans now wished to alter the common views in which China was understood as a colonial object managed by European empires or heavily controlled by the Japanese, allowing Americans to tap its markets directly and on equal terms.[39]

Scales, Centers, and Margins: Image as Map

By the final quarter of the century, picture puzzles were manufactured in scores by American puzzle makers.[40] As mentioned previously, by this time many of these toy manufacturers had incorporated themes from the American tableau—adding various specifically American topics, such as westward expansion and its accompanying racial and gendered myths, as well as U.S. foreign relations and their geopolitical implications for the nation to the originally European takes on geography, royalty, and Bible stories—as topics likely to entertain a wider demographic slice of American children. To cite an example, *The Prairie Scout* and the forty-piece "matched-picture" *Wild West Show* (1889), both manufactured by Milton Bradley and Co., adopted themes from their namesake shows, which had gained national and international fame during the 1890s.[41] These and similar other puzzles focused on the particularities of life in the oft-mythologized American West and were marketed mainly to American boys. Sold at fifteen and twenty-five cents respectively, they had interchangeable pieces that promised "a multitude of striking and attractive combinations" with which American boys were invited to imagine America's West, Native Americans, and frontier escapades as sites of aggressively masculine, playfully colonial alfresco adventure.[42]

Both nationalizing and domesticating child play at the same time, such seemingly innocent, imaginative sites of entertainment reproduced the racial and gendered complexities that postemancipation with which U.S. society was fraught. On the one hand, to the extent that their family's wealth allowed children access to mass-produced playthings, and following the financial whims of an ever-expanding children's market, postbellum toys and entertainment for children and youth typically addressed either boys or girls, and only rarely children in general.[43] This gendered division, which influenced both families and toy makers, was predicated on the notion of "separate spheres" and had roots in the many strict binary stratifications constructed and upheld in the United States, and in the West in general, between private and public, men and women, adults and children, and, consequently, between boys and girls. What is more, the masculinization of Wild West picture puzzles, though seemingly at odds with the feminization of the domestic U.S. empire that characterized much of the discourse on expansionism during the second half of the nineteenth century, suggests that dominant gender constructs, known as "gender order," surfaced in and affected different features of expansionist logic differently.[44] In the first place, for much of the century, the so-called civilizing and Christianizing missions were rhetorically bracketed as domestic acts of feminine benevolence. Indeed, the discourses and practices of establishing a domestic American empire were notably feminized during the nineteenth century. As Sánchez-Eppler points out, these discourses addressed and symbolized the white American Christian child.[45] Even so, territorial expansion and confrontational adventurism had long resonated with masculinity—a declaration of roles for an American public overly engaged with and invested in the rhetoric of expansionism. Wild West puzzles best reflected the convoluted gender order of the domestic U.S. empire as simultaneously masculine (hence uncompromisingly aggressive), feminine (therefore driven by compassion), and white (hence ordained by God and unquestionable). Put differently, the burgeoning global U.S. empire, primarily understood in terms of an outward-looking masculine project in the form of westward expansion, was also commonly symbolized by a carefully feminized view of domestication and civilization—that is, the cult of "manifest domesticity."[46]

On the other hand, Wild West picture puzzles transfused white America's century-long territorial quest to expand to the west with whitewashed racial modalities hard at work to define the relationship between white Americans of all classes and Native Americans across the continent. With the signing of new treaties and their breaching not long afterward, white and Native Americans were imagined by white America as benevolently interacting with each other in a long-running, graceful westward chase—otherwise celebrated by whites as the triumph of the expansionist American spirit over the western wilderness. The physical reality of the so-called westward movement at the heart of this chase fatefully, if convulsively, relegated Native Americans to the margins of the nation's territorial expanse. At the same time, this showcased the unapologetic whiteness and the dominant and toxic masculinity of the expansionist projects that had been mythologized as a defining trait of nineteenth-century white America.[47] In sum, by adopting scenes from Wild West shows, puzzle and game designers continued to emphasize the imperial masculinity and racial order of pioneer adventurism and later of off-shore expansionist projects, in as seemingly innocent and apolitical a material site of entertainment as that of children's playthings.[48]

Another example of geopolitically minded picture puzzles that brought a broader context to the play-study of the map of the United States is *A New Dissected Map of the World with a Picture Puzzle of the Capitol at Washington* (ca. 1890) by McLoughlin Brothers (figure 2.6). A two-sided puzzle, it was a dissected map on one side and a picture puzzle on the other. The puzzle's world map side includes the dissected map of the world with two clearly outlined boxes at the bottom—one showcasing a map of the United States and one containing a map of the British Isles. Stepping beyond the common cartographic practice of the time, the map side of this puzzle is in fact a collage of three maps drawn with three vastly different scales, all squeezed into one frame. It is striking to note that while the title reads "MAP OF THE WORLD" in large black font, much visual weight has been put on the importance of piecing together the United States and the British Isles as privileged, outstanding units of that world, all of which children were asked to piece together as "reinscriptive cartographers."

The image side of this puzzle adds yet another visual effect to the map collage side of the puzzle. While the map side maximizes the political and cultural proximity and the supposed global prominence of both a European empire and a young, aspiring republic, the image side displays a Victorian-style illustration of the Capitol in Washington in a serene and colorful setting. The visual effect is that the Capitol and the "Map of the World" are two sides

FIGURE 2.6 *A New Dissected Map of the World with a Picture Puzzle of the Capitol at Washington*, McLoughlin Brothers, ca. 1890 (The George Washington University Museum, Washington, D.C., AS 632 A, The Albert H. Small Washingtoniana Collection).

of the same puzzle, with the same number of pieces to make each of them whole. At the same time, while they are two sides of the same puzzle, it is fundamentally impossible for them to coincide visually; one can be seen only when the other is invisible. Moreover, with its two sides drawn at incomparably different scales, the puzzle interweaves the significance of politics on a global level with the exigencies of geography on a national level. From the perspective of its American manufacturers, one side deploys the smallest scale—in proportion to the physical size of the map—to include the entire world within its confines, while the other side zooms in to show the ideal seat of democracy and civilization as the materialized sum and idolized hope of that same world.

Next to classic geographical maps such as the *New Dissected Map of the World*, American publishers for children were also quick enough to pick up on current events and topics of the day and turn them into sensational, albeit noncartographic, picture puzzles. Alongside the dissected map of the Philippines mentioned earlier, picture puzzles such as *The Battle of Manila* (ca. 1898) and *Up the Heights of San Juan* (ca. 1898) by McLoughlin Brothers dissected scenes from the Spanish-Cuban-Philippine-American War.[49] *Up the*

FIGURE 2.7 *Up the Heights of San Juan: Our Boys Storming the Blockhouse in Front of Santiago*, McLoughlin Brothers, ca. 1898 (courtesy of Bob Amstrong's Old Jigsaw Puzzles).

Heights of San Juan, for one, depicted the Battle of San Juan Heights, fought on July 1, 1898, on the prominent points of San Juan Hill and Kettle Hill, east of Santiago de Cuba (figure 2.7). The battle was fought and won against the Spanish army by white and Black American soldiers, including both Theodore Roosevelt's Rough Riders of the First U.S. Volunteer Cavalry and the Black troops of the Ninth and Tenth U.S. Cavalry. The picture puzzle, though not a dissected map of Santiago de Cuba and its surroundings, was published almost immediately after news of the battle made it to the American press. *Up the Heights of San Juan* shows only the white American "boys" marching valorously forward against an enemy that is excluded from the illustration. But it is not only the enemy that finds no place in the illustration; similarly, African American soldiers who fought in the battle alongside their white fellow soldiers appear to have been hand-picked out of the rather crowded, exclusively white scene.

The absences and presences in this F. D. Maher illustration about such a strategic and celebrated battle depict the events as recounted by Theodore Roosevelt. Early in 1899, Roosevelt wrote a rather long memoir, serialized in *Scribner's Magazine*, in which he recounted several scenes from this consequential battle for the war-frenzied American public. Written by its most celebrated hero, this account of the war focused predominantly on white

American soldiers and left Black soldiers and the enemy forces out of the picture. Commenting on confusion in the ranks of American soldiers at the Battle of San Juan, Roosevelt wrote provocatively:

> None of the white regulars or Rough Riders showed the slightest sign of weakening; but under the strain the colored infantry-men . . . began to get a little uneasy and to drift to the rear, either helping wounded men, or saying that they wished to find their own regiments. This I could not allow, as it was depleting my line, so I jumped up, and walking a few yards to the rear, drew my revolver, halted the retreating soldiers, and called out to them that I appreciated the gallantry with which they had fought and would be sorry to hurt them, but that I should shoot the first man who, on any pretence whatever, went to the rear.[50]

Roosevelt's account of the battle was not left unanswered by the angered Black community. In fact, in response to Roosevelt's remarks, a letter to the editor was published in the *New York Age* in April 1899. Penned by Presley Holliday, a veteran of the Spanish-Cuban-Philippine-American War from the Tenth Cavalry, and directly in response to the press craze over Roosevelt's words, the letter invoked fellow soldiers and officers of all races, those "who can and will write history," those who can and will "release themselves from their voluntary status of military lockjaw," to write "tales of those Cuban battles that have never been told outside the tent and barrack room, tales that it will not be agreeable for some of them [Roosevelt and his proponents] to hear."[51]

Even so, Roosevelt's account shaped the image the public had of that battle to the extent that all the puzzles that emerged as popular sites for relating to what U.S. ambassador John Hay termed "a splendid little war" (including not only *Up the Heights of San Juan* but also the two other full-color picture puzzles manufactured by McLoughlin Brothers in the late 1890s, namely *Battle of Manila Picture Puzzle* and *The Rough Riders Scroll Puzzle*) banished African American, Spanish, and Cuban soldiers from the permissible boundaries of Americans' imagination of the conflict. These puzzles, and especially *Up the Heights of San Juan*, zoomed in as much as possible to include only those actors of the story that Roosevelt had included. The visual in political history "draw[s] boundaries of the political by deciding about the visibility or invisibility of topics, actors, or events."[52] In this vein, although not a dissected map in the classic definition of the term, *Up the Heights of San Juan* could be viewed as an unusually zoomed-in, political map, a pre-Google "street view" of the battlefield, which captured the events on the ground in a decisive battle that had been fought between one old and one emerging empire in one of the former's colonies, but neglected to represent their respective forces on the "periphery."

While zooming in to exclude other actors from the scene, laboring to present geopolitical relations in a figurative rather than a cartographic convention, *Up the Heights of San Juan* may still be understood as a map or an approximation thereof for a number of reasons. First of all, the picture puzzle's title pinned its topic to an exact point on the map of the world with known longitude and latitude. The San Juan Hill had an external referent both on the map of the world and on the mental maps of Americans, young and old, who were following the war news. Second, at a time when the "human turn" in geography as a discipline—itself at the service of and invested in world politics—had complicated the relationship between human beings, the environment, and larger political questions, this zoomed-in historic painting documented America's boot-on-the-ground, forward-looking presence in an international conflict and on foreign land. Third, by excluding Spanish and Cuban soldiers from the scene, the puzzle set a clear political border between Americans and their enemies, rendering visible the emerging U.S. empire's hinterlands on the global stage. *Up the Heights of San Juan* showed American soldiers as they marched forward from the western end of the picture to the east, mirroring the visual distribution of cardinal directions (north at the top, west to the left, and so on) commonly used in maps. Whether it was meant to be understood as a map and whether it was later viewed as an instance of cartography or merely as wartime memorabilia, *Up the Heights of San Juan* had one historically significant, spatial edge to it: it boldly and willfully positioned the United States at the center of an interimperial conflict that took place in the colonial margins of a waning European empire such as Spain, while at the same time heralding the United States' imminent march toward the center of the global imperial scene in the coming decades.

As yet another geopolitically inspired picture puzzle, and with 370 pieces, *The Nations at Peace* (figure 2.8) displays a striking thematic variation on the illustration that appeared on the 1893 cover of the *Clemens' Silent Teacher* previously discussed (see figure 2.5). While the *Silent Teacher*'s box-lid illustration showed a cheerful young white American male distributing cartographic representations of the United States among representatives of European empires and non-European nations, *The Nations at Peace* (ca. 1909) placed Uncle Sam—tall, old, and benevolent—at the center of a circle of ceremoniously dressed boys and girls holding hands. In this puzzle, each child stands for a

FIGURE 2.8 *The Nations at Peace*, Emily Batterson, ca. 1909 (courtesy of Bob Armstrong's Old Jigsaw Puzzles).

European nation, cheerfully and trustingly looking up to Uncle Sam. As the Belgian child, the youngest in the picture, grabs Uncle Sam's right knee, other children dance merrily around him, signifying the self-assigned role the United States assumed as an intermediary in the geopolitically unsteady Europe of the early twentieth century. This illustration seems to testify that during the years between the closing of the western frontier and the outbreak of World War I, that same young American male aged into wisdom by coming out of the Spanish-Cuban-Philippine-American War victoriously, so much so that he was now ready to assume the role of a buoyant international arbiter.[53] With increasing tensions in American international relations following the Spanish-Cuban-Philippine-American War, the rival German and Japanese world powers on the rise, the ongoing conflict between France and Germany over Morocco, and the Russo-Japanese War during the first decade of the new century, the early years of the twentieth century tested the young world power's forte in foreign policy. By the time *The Nations at Peace* was copyrighted, Elihu Root—President Roosevelt's secretary of state and former secretary of war under President McKinley—had signed five-year arbitration treaties with twenty-four nations, including France, Great Britain, Spain, Norway, Sweden, Japan, Costa Rica, Portugal, and Italy.[54]

American public school curricula, too, reacted to the new, internationally engaged role the U.S. government was performing. In 1906, and for the first time since its inception over seven decades earlier, the American Institute of Instruction, for instance, initiated a Department of Peace and held a meeting on peace education in American public schools. Addresses were made at the meeting with topics such as "Educating People for International Arbitration" by William H. P. Faunce (1859–1930), the president of Brown University, a critique of jingoistic and superficial false friends of patriotism by Arthur M. Wheeler, a talk that addressed "the unfortunateness of the enormous naval display ... and expenditures on armaments" by Mrs. Mead, and suggestions to adopt notions of peace and war in school curricula.[55] Reflecting on the shifting international waters and the U.S. government's involvement in overseas conflicts, including in its own colonial advances in the Pacific and the Caribbean, the meeting's report summarized its resolutions as follows: "that all teaching should be permeated by the peace spirit, and that a committee be appointed to prepare a plan for organizing the teachers of the country for an active campaign of peace instruction in the schools."[56] *Nations at Peace*, illustrated by an unknown artist and dissected by Emily Batterson in 1909, most probably refers to a similar position that many American pacifists of the time, including educators and teachers, took with regard to the nontraditional role the United States had begun to assume in its foreign policy by the start of the twentieth century.

By the end of the century, it appears as though the entertaining side of dissected maps and maplike picture puzzles had surpassed their educational function; this is evident in the way toy makers began to further diversify the themes that they chose for their puzzles. As an example, the picture puzzles *Fire Engine* and *Department Stores* put emphasis on entertainment and everyday urban life and addressed children of both sexes. In the meantime, patriotic themes such as the expansion of the U.S. Navy (such as F. D. Maher's *The North Atlantic Squadron Picture Puzzle*, produced by McLoughlin Brothers ca. 1898), were selected for dissection in ever-larger sizes and into ever more pieces—still following the principles that mattered to the nation, both domestically and internationally, at the dawn of a new century. Beyond doubt, the shifting overseas interests and geopolitical imperatives of the nation had turned American puzzles into unique sites of reproduction, appropriation, and further manipulation of common cartographic representations, both of the world and of any smaller parts of it that were selected for dissection.

Taking a step back to view this phenomenon within more comprehensive theories of cartography, there is overwhelming consensus that the geographic map is selective in its topic of choice and the scales used to draw it, and that it entails intentional absences and presences. As Gilles Deleuze and Felix Guattari have pointed out, mapping thus involves "an experimentation in contact with the real." In its wake, mapping further "constructs the unconscious" that is involved in this not necessarily one-to-one representation.[57] According to Deleuze and Guattari, unlike simple acts of tracing, cartography includes connections, associations, interests, disjunctions, and paths well beyond the surface of the region, city, country, or continent that it represents. This is exactly how a dissected map works, too. As this chapter amply illustrates, in the case of turn-of-the-century dissected maps, it was not only the scales used in the original map that mattered to the United States in geopolitical terms; the sociopolitical imperatives of the time could and did take regions that had not formerly been considered important enough to be dissected and turn them into desirable wholes ready for dissection. This practice logged these newly important locations in an "unconscious directory," which Deleuze and Guattari point to in their definition of the map beyond its representative function.[58] Decisions involved in designing and producing dissected maps—(dis)regarding political borders; following subtle politics of coloring and of renaming colonized regions; and selecting accompanying maps, box-lid images, or illustrations on the flip side of two-sided dissected maps—are all instances of how such a cartographic "unconscious" was constructed for children who played with and learned from dissected maps. Dissecting maps not only brought spatial realities within the sensory reach of those who manufactured and those who played with them; more than that, it involved selecting, dissecting, playing with, and consuming such a reality based on geographic conventions, school geography lessons, children's creative patterns of absorption alongside their occasional boredom with geography, and American adults' time-bound and time-specific understandings of the world and their place in it.[59]

CLOSELY FOLLOWING THE BIRTH and the century-long evolution of a national school geography curriculum in the United States, I have based my arguments in chapters 1 and 2 on the premise that at the turn of the twentieth century, a clear majority of Americans—and some classes doubtlessly more than others—were drawn to, and benefited from, the expansionist tendencies that connected the United States to the world both commercially and diplomatically. Maps, though not yet at their peak of appeal to the general public, as they would be during World War II, served as proxy *Treffpunkte* that allowed the rising middle-class America "to imagine and comprehend a world that most would not comprehend firsthand."[60] The turn of the twentieth century, consequently, should be understood in hindsight as a moment at which mental, physical, and political maps inevitably competed against one another—each functioning as a proxy for the others, and each veiling realities of spatial disparity while encouraging other interpretive possibilities.

What is more, as chapters 1 and 2 illustrate, it is a folly to overlook the fact that children consume childhood—including school primers, extracurricular reading material, and their toys—in both compliant and subversive ways. Children, as Sánchez-Eppler remarks, "transmute compliance into play."[61] It is true that American children would try to memorize lists of place names that appeared in their geography primers, but they would also forget some of them, mistake one for the other, or have difficulty pronouncing or spelling them for reasons ranging from laziness to playfulness, from lack of trust in the geography teacher to a preference for subjects such as arts or arithmetic. As chapters 3 and 4 illustrate, at the intersection of these and other unaccounted-for factors, children consumed school geography playfully, reminding adults that imagination and didactics, transgressions and misappropriation, are not only concurrent with one another but also coterminous with childhood.[62]

Likewise, childhood continues to be a far more complex, fragile, and reconstructible artifact when lived by actual children. Consequently, although childhood was, and to a large degree still is, an adult-made construct, there is no denying that children have had their own patterns of consuming, deconstructing, and repurposing its many facets, including geographical education and the playthings intended to foster it. As chapters 3 and 4 make the case, by implanting child protagonists, playthings, and playmates into letters and geographical puzzles; playfully adopting synecdoche and homonymy while bringing politically charged colonial place names into their apolitical micronarrative story-puzzles and their seemingly well-ordered alphabetic puzzles; juxtaposing the American home with an assortment of homes (and non-homes) across the globe (even within the colonies); and acting as heedless participants in colonial undertakings (roles that exceeded their ambivalent, pre-political, and seemingly innocent subject positions), children troubled the binaries of colonizer–colonized, U.S. empire–the world, and American–non-American.

CHAPTER THREE

A for Amoy, Z for Zanesville
Child-Made Geographical Puzzles, Finger-Tip Travelers, and Cartographic Intimacies of the World

Children's simultaneous assumption of a predominantly ambivalent position as pre-political subjects and their unfailingly persistent and frequently playful engagement with questions of territoriality, national identity, and geopolitics can be traced in the textual evidence they left behind in writing. The first of the two sets of child-authored geography-themed texts that I look at in this book—that is, puzzles printed in *Harper's Young People* during the 1890s—is the subject of this chapter.[1] Reading children's letter-writing and puzzle-composing activities as appropriating cartographic matter and navigating space, my purpose here is both to identify some of the ways American children imagined, described, and recycled their knowledge of the world as it related to the United States and to study how they rejoiced in the possibility of repurposing formal geographic literacy by writing themselves into the nation's global imperial narrative.

Without prior notice, the practice of printing puzzles by "young contributors" started in the fifth issue of *Harper's Young People*, dated December 2, 1879. In response to George S., a young reader who had asked about sending puzzles to the weekly, the editor wrote: "We will accept original puzzles if they are very good. They must, in all cases, be accompanied by a full solution."[2] In a matter of weeks, Kate S., nine years old, sent *Harper's Young People* some puzzles, which, though not ultimately printed, were praised by the editor in the December 30 issue as "very neat for such a little girl to compose." These letters initiated a tradition that was to last for more than a decade. As these instances indicate, it seems that *Harper's Young People*'s editors were at first surprised by the interest children expressed in participating in their favorite weekly not only as readers but also as contributors. In later issues, the editors explained that originality was the main criterion for a puzzle to make it to the magazine's pages. But other elements, too—such as unnecessarily excessive praise for the magazine, inclusion of the puzzle maker's full name and address, and the difficulty level—would also influence the editors' decision to print child-composed puzzles. For example, a geographical puzzle sent by Martha W. D. to *Harper's Young People* could not be printed because "we are afraid our young readers would never make it out, as it requires an extraordinary amount of geographical knowledge."[3]

Soon eight puzzles by children of different ages and genders were printed in the eleventh issue of the magazine.[4] Following this trend, in its February 17, 1880, issue, *Harper's* launched a regular column, called Puzzles from Young Contributors, as a space to print children's puzzles.[5] After that, puzzles were usually printed right after children's letters, as the very last column in the weekly before the advertisements and the back cover. As an unwritten rule, they were usually printed without disclosing their composers' full names or ages. Correct answers to each puzzle were printed two or three issues later, together with a list of the names of those children who had answered all or some of the puzzles correctly. By 1880, almost every issue of *Harper's Young People* printed at least one geographical puzzle. In fact, at the start of the following decade, almost every issue of the weekly included at least one—and usually two or three—geography-themed puzzle of different types and varying difficulty levels. Over time, more geographical puzzles were printed, some of which asked about fairly obscure toponyms. By the *Harper's Young People's* fifteenth issue, as more and more puzzles came in, the editors asked children to make sure, in the case of geographical puzzles, not to "take a name little known, like that of some Western town, to form an enigma, for children in some other part of the country will find it difficult to solve."[6] Throughout the following eleven years, however, the need to remain original meant that more and more obscure toponyms—especially of non-American places—found their way into geography-themed puzzles.

Child-authored geographical puzzles varied in type. Some were micronarratives characterized under the heading of "Geographical Guesswhat" or "Geographical Guessthestory." In each story, some words were omitted, including proper names or the names of various animals, foods, and so on. In order to complete these seemingly nongeographical stories, however, children had to find out the natural features or the place names that the story referenced. Others—listed under headings like "River Puzzle," "Alphabetical Cities," and "Geographical Puzzle"—consisted of alphabetically ordered riddles about various natural features or place names from all over the world. Most commonly, these alphabetical lists asked for place names starting with the letters of the roman alphabet, usually all twenty-six of them. Less frequently printed were the more complicated verbal enigmas characterized as "Pied Cities," "Pied Islands," or "Hidden Rivers," which asked for more obscure geographical names in a cryptic way. These geographical enigmas scrambled the letters of the alphabet within a word or hid a place name's letters in the correct order within two or three adjacent words in a random sentence and asked puzzle solvers to figure out the name. In designing and then solving these puzzles, children had their own means of exerting power over the archives they were inadvertently building by ensuring that they composed puzzles that were original, which would at times make them nearly impossible to solve.

Invested in the collision of semantic charge and spatial meaning in children's attempts at bonding certain place names with individual letters of the alphabet, the present chapter surveys these puzzles, published by *Harper's Young People* during the 1890s, with a particular interest in the territories and letters of the alphabet that children invoked most frequently in their intimately homonymic, idiosyncratically encyclopedic, or eclectically synecdochical references to world geography—that is, the cities, regions, watersheds, and larger geographical entities that held their attention on the world map. Looking beyond these puzzles' entertaining and showy façades, my aim is to trace their patterns of reference as they suggest historically key associations with and dissociations from landscapes and civilizations, stereotypical preferences for certain places and regions over others, and scalar confusion and tension in the young puzzle makers' geographic knowledge of the world that lay beyond, and yet included, their U.S. home. To provide a framework for understanding the models and forms that influenced child puzzle makers, I start the discussion with a brief examination of geographical puzzles that were composed, or at times in fact physically manufactured, by American adults at the turn of the twentieth century, highlighting what they did to and with the planet's territoriality. The chapter then proceeds to examine two popular forms of children's geographical puzzles. A first analytical section examines patterns of reference to various world regions in short-story puzzles as well as the homonymous relationship between place names and proper names; this is followed in the second section by an examination of alphabet-based puzzles and the synecdochical and encyclopedic relationship between letters of the alphabet and place names in these puzzles.

Before delving into a closer examination of these puzzles and what they tell us about children's liaison with empire, it is vital to place the forthcoming discussions in this chapter in the broader conversations on where children stood in the culture market of the time. First, we should keep in mind that for a great majority of these more or less untraveled children, major parts of the United States and the entirety of the non-American world—including the so-called Orient and Europe (and its Other in the south, Africa)—could at best be experienced as a territory of imagination. Acts of (re-)naming, which Rebecca Ann Bach understands as the simultaneous "naming and transforming

[of] space,"[7] point at the tangle of relations between children and the non-American world. Echoing Said's classic understanding of "the Orient" as an equally unfocused and slanted term that is engrossed in a "field of meanings, associations, and connotations"—children both doctored the surface of the earth with toponyms and injected it with the many layers of intimate, but also violent, connotations that those toponyms carried with them.[8] Upon consuming adult-generated school geography, works of fiction, travelogues, and geographic games and toys, children had the chance, my argument goes, to parse the geographic knowledge and cartographic data they had received at school and to work with what had interested them and what they found important with respect to their immediate, localized lives. Although they present us with partial and shifting evidence, the puzzles they composed put forward a sketch of the particular understanding of the world that young, geography-savvy citizens of the globalizing U.S. empire were developing at the turn of the century.

Second, and in a broader context, we should attend to the contradictions that unfold in the cross-fertilization between formal education, children's play, adults' political agendas, and popular media platforms that takes place in this accidental repository of child authors and creators who had sparse publishing opportunities. On the one hand, children's attempts at composing printable puzzles (as well as printable letters that are under study in chapter 4) constitutes an exciting, if overlooked, element of a culture that welcomed children's participation in a competitive marketplace, allowing them a space of playful, reward-based authorship. On the other hand, it should not be forgotten that *Harper's Young People*'s Puzzles from Young Contributors was heavily structured and always tucked at the seemingly unadventurous end of a weekly that, though *for* children, was not mainly *by* children. As a manifestation of this division between intended readers (children) and the majority of the magazine's contributors (adults), children's letters and puzzles were comfortably hidden away at the back of the magazine, a nondisruptive way of enacting an editorial policy that made the weekly appear welcoming of children's voices.

Puzzles by American Adults: World Geography and Geopolitics

Puzzles as a pastime have enjoyed a long history of steady growth and consistent public interest regardless of language and geography. And yet it was only late in the nineteenth century that puzzle making became a lucrative line of business on either side of the Atlantic. By this time, professional puzzle makers emerged in England, France, and the United States, and associations such as the U.S.-based National Puzzlers' League were founded.[9] Samuel Loyd, the most celebrated American puzzle maker of the time, was able to live a comfortable life based on his skills as a puzzle maker and an ingenious promoter of his products. During the 1890s, Loyd's column on puzzles for the *Brooklyn Daily Eagle* was among the most popular daily columns in the country.[10] In 1897, together with his son Walther, Sam Loyd began to independently publish his own puzzles in the form of a monthly publication called *Our Puzzle Magazine*. In this chapter, I analyze Loyd's enormously popular creations in order to examine the geographic concepts and register the geopolitical concerns and values that adult American puzzle makers relied on as the common knowledge that would enable the public to enjoy solving their puzzles.

In 1896, and toward the end of the economic depression known as the Panic of 1893, Loyd lent his puzzle-making skills to the Republican Party's presidential campaign (figure 3.1).[11] Designing the mechanical puzzle called *Get Off the Earth*, Loyd in fact helped President McKinley win a cutthroat presidential election against Democratic candidate William Jennings Bryan.[12] Well over a million copies of this "Puzzle Mystery" were distributed at the time as a free giveaway to promote McKinley's unavoidably home-based campaign for reelection.[13] The mechanically innovative, egregiously racist board puzzle showed thirteen Chinese warrior figures on a rotating globe and asked people to try to decide in which direction the disk should be turned in order to make one of the warriors disappear and to explain how they think this happens.[14] At a time when the Chinese Exclusion Act of 1882 was still very much in effect, and while the United States sought to negotiate an open-door policy toward the massive Chinese market, the lion's share of which was already controlled by four European empires, *Get Off the Earth* served the Republican Party's agenda as a clear yet playful reference to "the Chinese question" as both a domestic and a foreign policy issue for the United States. A year later, and still cashing in on the influx of non-European, nonwhite migrants to the United States as a mere question of numbers, Loyd copyrighted another edition of the same puzzle called *The Lost "Jap."*[15] Almost one million copies of this puzzle, the type of which Marcel Danesi refers to as "disappearance puzzles," were distributed by the Metropolitan Life Insurance Company (figure 3.2 (a))[16]. The company offered twenty prizes, ranging from five to a hundred dollars, for the best explanation of the puzzle's mechanism within one year of the puzzle's distribution.[17] In the case of both puzzles, and in reinscribing the national as the global, the political as the

FIGURE 3.1 Advertisement of *Get Off the Earth Puzzle Mystery*, Sam Loyd, 1896 (courtesy of Lilly Library, Indiana University, Bloomington, Indiana).

private, and the demographic as the geographical, the quest is flagrantly necropolitical: How could the U.S. public join hands with politicians in proto-global efforts to decimate the displaced Chinese and Japanese populations across the globe (compare figures 3.1 and 3.2) even before they reached the United States as undesirable bodies that needed to be quarantined, regulated, sanitized, and immobilized?[18]

These puzzles were followed by the *Puzzle of Teddy and the Lion* in 1909. This third disappearance puzzle outlines the struggle between eight men (Theodore Roosevelt and seven African men) and seven lions and asks the players to find *Which Black Man Turns into a Yellow Lion?* (figure 3.2 (b)). Hinting at the carnivorous nature of lions, the puzzle equated African men to savage lions, relying on a slightly different coded message than that in the case of

FIGURE 3.2 (A) *The Lost "Jap,"* Sam Loyd, 1897 (courtesy Lilly Library, Indiana University, Bloomington, Indiana).

FIGURE 3.2 (B) *Puzzle of Teddy and the Lion*, Sam Loyd, 1909 (courtesy Lilly Library, Indiana University, Bloomington, Indiana).

the Far Eastern warriors of *The Lost "Jap."* In effect, the *Puzzle of Teddy and the Lion* highlighted the thin line between the "hominine feral" African (that is, according to then-current racist theories of "civilizational hierarchy," a nearly "civilized" human currently regressing to a lesser stage of civilization) and the fully animalistic *feline* (that is, the African lion). While the racially stereotyped African men are all marked as potential lions (or as their prey), what remains intact in this puzzle are planet Earth and the purportedly adventurous, masculine white American hero, Teddy Roosevelt.

Furthermore, we can read the puzzle in line with attempts to address the so-called crisis of masculinity that had followed the symbolic closing of the western frontier in the 1890s by molding American masculinity in the figure of Western pioneers and the Rough Riders, and, especially, in the figure of Theodore Roosevelt.[19] As now classic studies by Gail Bederman, Michael S. Kimmel, Kristin L. Hoganson, and Amy S. Greenberg, among others, showcase, by the end of the Reconstruction era and with the closing of the western frontier, white Americans responded to new waves of what they perceived, discursively, as a not-fit-for-empire emasculation of white American boys and men. In response, and pioneered by turn-of-the-century theorists, policy makers, and educators who believed in the necessity of revising and recalibrating white American masculinity, rough, outdoorsy models of masculinity were promoted (pioneers, Boy Scouts, and the Rough Riders, for example) both to "save" the nation's youths and to ensure the triumph of the white established classes over nonwhite immigrants and formerly enslaved Americans and their descendants.

Puzzles deal with more than human curiosity; they entail microlevel efforts by various groups of people to create diversion, to find answers to macrolevel crises, and to address global questions. As Loyd's disappearance puzzles illustrate, various types of puzzles have historically offered people the opportunity to seek meaning when they feared its absence by, in Danesi's words, "providing small-scale experiences of the large-scale questions that life poses."[20] What Loyd's disappearance puzzles celebrated was that as Western "civilization" was marching in all directions and as white Americans were solidifying into a nation that supposedly stood against the wonders of nature on a global scale, human "races" (other than the "enlightened" and the supposedly outnumbered white, masculine Americans) could purportedly be reduced in number to make space for or guarantee the safety of the adventurous, righteous, globe-trotting white Americans. Loyd's geographical puzzles thus created imaginary cut scenes that depicted ways the American public could partake in U.S. foreign policy, in the so-called civilizing mission, in the expansionist projects of their rising overseas empire, and in the regulation of U.S. borders in the face of new waves of immigrants.

Puzzles by American Children: Cartographic Intimacies of Empire

While Loyd was most famous for his mathematical puzzles and chess problems throughout the 1870s and 1880s on both sides of the Atlantic, his politically informed and racist disappearance puzzles proved to be among the most successful and popular of his inventions.²¹ As we saw, these mechanical puzzles linked themes from U.S. domestic and foreign policy with populist ideas about anti-blackness and anti-Asian racism, masculinity, geopolitics, and the "civilizing" mission. Against this backdrop, American children, too, felt encouraged to come up with all sorts of puzzles—algebraic, mathematical, and geographical—and send them to such popular magazines as *Harper's Young People*.²² The discussions that follow examine two broad categories of puzzles by young composers: short-story puzzles and alphabetical puzzles. Here I demonstrate that young puzzlers composed their puzzles by building an intimate, yet not necessarily accurate or coeval relationship to the place names they asked for as solutions to their puzzles. Symptomatic of children's overall preoccupation with their local lives and away from the geopolitical concerns that adult-authored puzzles favored, this intimate inaccuracy emerges in at least two ways: First, through the homonymy between their own or their friends' and siblings' given names and those of recently colonized places on the map of the world, and second, by establishing a synecdochical—that is, *part-for-whole*—relationship between letters of the alphabet and place names on the world map.

Short-Story Puzzles and Homonymy

The Geographical Guessthestory puzzle printed in *Harper's Young People*'s September 9, 1890, issue is in the first place a short story. As the story starts, we learn that an American boy named *island in the Indian Ocean* lives in "a very beautiful home" in Philadelphia with his parents and his two sisters, *town in Ohio* and *town in Idaho*. One day, he, his sisters, and his cousins—*town in Alabama, two capes east of Virginia,* and *first half of town in Illinois*—go on an adventure in the woods near their home, build a fire and make coffee, have lunch, and talk about their many pets (figure 3.3). As expected of well-mannered children, they return home before sunset. Nothing adventurous happens on this uneventful afternoon during which Alphonse, Marion, and Florence are not reported even to have played hide-and-seek with their cousins in the countryside. Despite the story's insistence on presenting a moral in the end—as a typical nineteenth-century children's story written by an adult would most probably do—there can be little doubt that the puzzle maker is a child, because while the story mentions that he, his two sisters, and his six cousins go on a picnic, its unsigned composer manages to name only four of the six cousins throughout the rest of the narrative, and this remains unedited in the printed version of the story.

This story puzzle is a simple and unadventurous account of a children's picnic lunch in the woods somewhere along the East Coast of the United

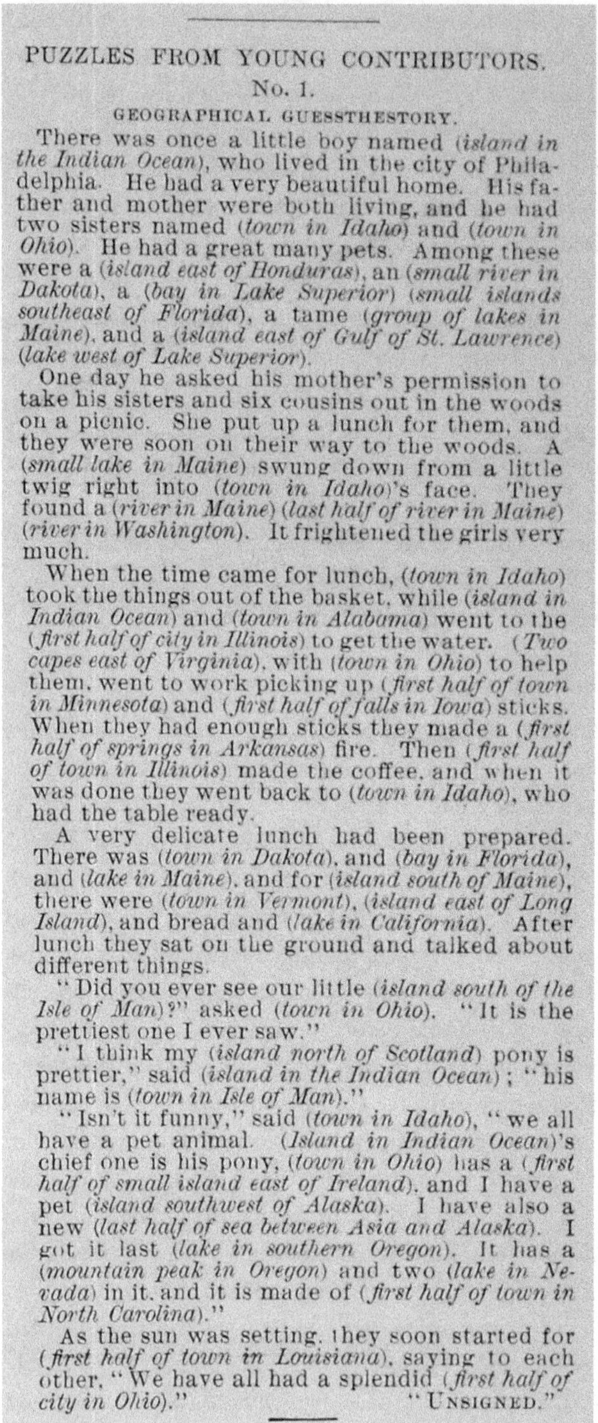

PUZZLES FROM YOUNG CONTRIBUTORS.
No. 1.
GEOGRAPHICAL GUESSTHESTORY.

There was once a little boy named *(island in the Indian Ocean)*, who lived in the city of Philadelphia. He had a very beautiful home. His father and mother were both living, and he had two sisters named *(town in Idaho)* and *(town in Ohio)*. He had a great many pets. Among these were a *(island east of Honduras)*, an *(small river in Dakota)*, a *(bay in Lake Superior)* *(small islands southeast of Florida)*, a tame *(group of lakes in Maine)*, and a *(island east of Gulf of St. Lawrence)* *(lake west of Lake Superior)*.

One day he asked his mother's permission to take his sisters and six cousins out in the woods on a picnic. She put up a lunch for them, and they were soon on their way to the woods. A *(small lake in Maine)* swung down from a little twig right into *(town in Idaho)*'s face. They found a *(river in Maine)* *(last half of river in Maine)* *(river in Washington)*. It frightened the girls very much.

When the time came for lunch, *(town in Idaho)* took the things out of the basket, while *(island in Indian Ocean)* and *(town in Alabama)* went to the *(first half of city in Illinois)* to get the water. *(Two capes east of Virginia)*, with *(town in Ohio)* to help them, went to work picking up *(first half of town in Minnesota)* and *(first half of falls in Iowa)* sticks. When they had enough sticks they made a *(first half of springs in Arkansas)* fire. Then *(first half of town in Illinois)* made the coffee, and when it was done they went back to *(town in Idaho)*, who had the table ready.

A very delicate lunch had been prepared. There was *(town in Dakota)*, and *(bay in Florida)*, and *(lake in Maine)*, and for *(island south of Maine)*, there were *(town in Vermont)*, *(island east of Long Island)*, and bread and *(lake in California)*. After lunch they sat on the ground and talked about different things.

"Did you ever see our little *(island south of the Isle of Man)*?" asked *(town in Ohio)*. "It is the prettiest one I ever saw."

"I think my *(island north of Scotland)* pony is prettier," said *(island in the Indian Ocean)*; "his name is *(town in Isle of Man)*."

"Isn't it funny," said *(town in Idaho)*, "we all have a pet animal. *(Island in Indian Ocean)*'s chief one is his pony, *(town in Ohio)* has a *(first half of small island east of Ireland)*, and I have a pet *(island southwest of Alaska)*. I have also a new *(last half of sea between Asia and Alaska)*. I got it last *(lake in southern Oregon)*. It has a *(mountain peak in Oregon)* and two *(lake in Nevada)* in it, and it is made of *(first half of town in North Carolina)*."

As the sun was setting, they soon started for *(first half of town in Louisiana)*, saying to each other, "We have all had a splendid *(first half of city in Ohio)*."   "UNSIGNED."

FIGURE 3.3
"Geographical Guessthestory," *Harper's Young People*, September 9, 1890 (courtesy of Library of Congress).

States. However, in order for the young puzzle solvers to complete the story, they had to take a rather long and adventurous journey on the table globe, the wall map, or the home atlas in order to find the proper names of all the children, food, and pets mentioned in it. It is only Philadelphia—that is, the children's home—which is named in the text from the beginning. Other than that, in addition to references to lakes, rivers, waterfalls, bays, mountain peaks, towns, and islands across the United States, and especially on the East Coast, this otherwise unadventurous story reached as far as Scotland, the Isle of Man, Ireland, Newfoundland, Asia, Central America, and the Indian Ocean. In effect, in order to complete the story, children had to pore over maps or table globes and move about the surface of the earth in its accessibly scaled-down representations, just as world travelers would on the physical surface of the planet. The puzzle solvers were supposed to spot Alphonse east of Africa, to scale down this minuscule former French and, later, British possession in the Indian Ocean, to invoke homonymy in order to imaginatively personify that island as a little urban American boy, and finally to place it/him on the map of the United States as the story's protagonist. As in any (re)naming project, this was a genealogical journey that related bodies to spaces—a journey made possible not through generational cycles of birth and death but through eradication and erasure.

This puzzle was made by an American child who would probably have been unaware of, and uninterested in, the complex transoceanic colonial struggles between the local population of an isle such as Alphonse and its Arab, Indian, and later European possessors since at least the sixteenth century. Alphonse Island, located seven degrees south of the equator in the Indian Ocean, was "discovered" by the French navy commander Chevalier Alphonse di Pontevez in 1730. Earlier records suggest that the isle was marked as "San Francisco" by the Portuguese early in the sixteenth century. While the French rule over this and other islands northwest of Madagascar was contested by the British in the nineteenth century, the island continued to be a coconut plantation.[23] Disregarding the complex history behind the island's name, perhaps even considering it an entirely arbitrary geopolitically inconsequential piece of information within a geographical education built on memorization of names and figures, the puzzle composer reduced Alphonse to its European "given name" on a map that had been produced in the imperial cartography workshops of western Europe or the United States. The puzzle maker further took Alphonse out of its geographical context—exotic as an island off the east coast of Africa and yet familiar as an American boy's name—and established a new context for it by connecting it back to a person. However, while the island received its name in a colonial context when the original Alphonse renamed it after himself, the puzzle's Alphonse was a fictional boy living in urban America—a young citizen of the United States who was expected to prepare, along with others in that generation, to become a "steward" of the nascent overseas U.S. empire.

Given the politics and the layers of sensory as well as intellectual cognition involved in the process of composing and solving this and similar story puzzles, it is imperative to examine how they conflated common geographical and personal imaginings of the world in the close but infrequently visible negotiations between geographically privileged adults—namely colonizers, mapmakers, professional geographers, and *Harper's Young People* editors—and their young audiences. As previously noted, what makes this and similar puzzles worthy of in-depth examination are the spatial, pedagogical, political, and personal layers of composition that a select number of geographically savvy children brought together in seemingly trivial puzzles. The thematic investigation of these puzzles as a textual body of source material, beyond their modest facades, opens up possibilities to uncover the common understandings of world geography among adults and children and to identify where and how children's understandings diverged from those of the adults who instructed them.

In contrast to the spatial work done by Du Bois's commentary on the American breakfast table (which appeared in the discussion of home geography in chapter 1), in which he links the very interior of American homes to various points across a fragmently mapped globe, many children connected their idea of home to the world at large, stepping outside their homes in these globe-trotting story puzzles. Similar to the story puzzle just analyzed, Frank Rossman's River Puzzle, printed in the November 12, 1889 issue of *Harper's Young People*, is a child's account of the outdoor adventures of a group of American children. Frank's puzzle refers to a handful of inconsequential, unpleasant, yet adventurous things that could happen to children outside the home. Scene one: Two boys go for a walk in the countryside where they encounter a snake. Scene two: A friend of theirs reports that she was bitten by a venomous bug (figure 3.4). The story ends. This short, nearly plotless, cautionary "babes-in-the-wood" type of narrative is completed by inserting the names of rivers from around the world into its incomplete sentences.[24] Here, again, a homonymous relationship is established between river names and common Anglo-Saxon proper names (rivers in the United States, Scotland, and Australia); common American pet names (a river in South America); and generic animal names in English, such as "snake" (a river in the United States) and "bug" (a river in Poland), transforming the child-made puzzle—an artless

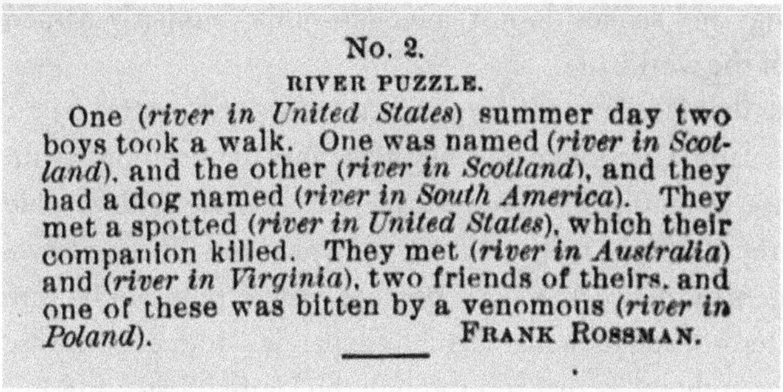

> No. 2.
> RIVER PUZZLE.
> One (*river in United States*) summer day two boys took a walk. One was named (*river in Scotland*), and the other (*river in Scotland*), and they had a dog named (*river in South America*). They met a spotted (*river in United States*), which their companion killed. They met (*river in Australia*) and (*river in Virginia*), two friends of theirs, and one of these was bitten by a venomous (*river in Poland*).
> FRANK ROSSMAN.

FIGURE 3.4 "River Puzzle," *Harper's Young People*, November 12, 1889 (courtesy of Library of Congress).

fantasy report on the daily lives of rural American children—into an oblique catalog of actual reference points on the map of the world.

The homonymous relationship that the puzzle establishes between rivers and people is anything but simple. On the one hand, children adopted homonymy in turning their stories into puzzles as a technique with which they imagined themselves and their country beyond the safe borders of home, placed in a world that American adults had promised would be adventurous yet familiar and tolerably safe.[25] For the untraveled young American child, this dual quality was, at least partly, the result of naming acts by various European empires, some of whose citizens had since chosen the United States as their home. Such a world, at least in these story puzzles, was named after American children, their parents, their siblings, or their pets. The world was thus made mappable, navigable, and homelike. While leaving out adult characters, such as parents, teachers, or servants, from their story puzzles (Alphonse's mother's strictly domesticated role, for example, was limited to preparing a "very delicate lunch" for their picnic [figure 3.3]), children benefited from homonymy in their imaginary, free-of-charge, unchaperoned travels to different parts of the world—exotic and yet familiar. While thus traveling, they remained ignorant of the violent, at times genocidal, actualities involved in the naming acts, by means of which generations of adults had aggressively pinned such "familiar" names onto the surface of the globe. Writing in the metaphoric hinterlands beyond the tangle of violence at the heart of adults' projects of claiming and renaming places, the child authors of these puzzles (but also those children who were addressed by them as readers of *Harper's Young People*) had the luxury of overlooking original local toponyms and unimagining the actual past and present inhabitants of these "dots" on the world map in order to further expand the borders of home—that indoor sanctuary where bugs and snakes do not enter—to other familiarly named places on the map of the world.

At the same time that it enlarges homonymous words' associative fields, homonymy also supplants a predominantly negative sense of uncertainty with a more neutral sense of ambiguity. Interlacing Frank's puzzle with double meanings, homonymy allowed American children to play outside or to travel to near and far corners of the world exactly because adults could not be sure whether what threatens the child in this sketchily expanded draft of home are bugs and snakes or merely rivers that run through the United States and Europe. Needless to say, adults knew that certain names not only pointed to specific human beings but also contained in themselves references to geographical spaces where colonial power games were played. Renaming has long been adopted as a potentially tentative and revocable yet irrefutably violent project of imperial appropriation and colonial reclamation of space—what Brückner identifies as "systematic eradication" of places and people—and as such, it both facilitates and justifies the spatial and cultural advances made by the colonizer.[26] The act of renaming alienates the colonized not only from their past and present, their language and culture, but also from their right to protest the depletion of the limited natural resources the colonized spaces have in store, the destructive impact of such depletion on Indigenous ways of life, and the ensuing ecological crises that will assume global dimensions in the decades and centuries to come. Therefore, on the face of it, to interlink places that held potential natural resources for the United States' rise to global prominence with the lives of young American children through the familiarity of proper names might seem like an innocuous occasion to further practice the tenets of a geographic curricula that centered around the idea of home geography. What this interlinking in fact accomplished, however, was to recalibrate and rewrite large-scale U.S. geopolitical investments in the world as an intimate feature of the immediate everyday lives of white American pupils.[27]

Overall, in revisiting and further complicating the notion of "home geography" discussed in chapter 1, I would like to insist that, informed by the teachings of turn-of-the-century professional geographers like William Morris Davis and Richard Elwood Dodge, who stipulated that geography be thought of and taught as a multitude of comparable home geographies across the globe, children understood world geography in composites, at once homebound *and* alfresco, local *and* global. For example, Dare and Frank K., whose puzzle was printed in the November 26, 1889, issue of *Harper's Young People* (figure 3.5), composed a narrative in which two brothers, *a lake in New York*

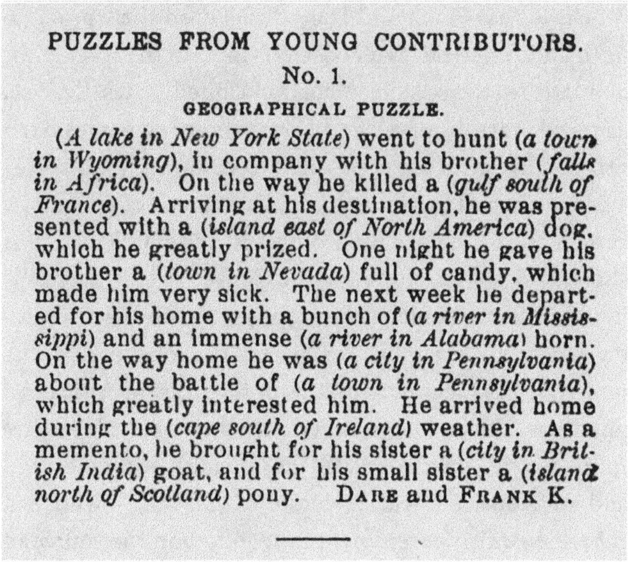

> **PUZZLES FROM YOUNG CONTRIBUTORS.**
> **No. 1.**
> **GEOGRAPHICAL PUZZLE.**
>
> (*A lake in New York State*) went to hunt (*a town in Wyoming*), in company with his brother (*falls in Africa*). On the way he killed a (*gulf south of France*). Arriving at his destination, he was presented with a (*island east of North America*) dog, which he greatly prized. One night he gave his brother a (*town in Nevada*) full of candy, which made him very sick. The next week he departed for his home with a bunch of (*a river in Mississippi*) and an immense (*a river in Alabama*) horn. On the way home he was (*a city in Pennsylvania*) about the battle of (*a town in Pennsylvania*), which greatly interested him. He arrived home during the (*cape south of Ireland*) weather. As a memento, he brought for his sister a (*city in British India*) goat, and for his small sister a (*island north of Scotland*) pony. DARE and FRANK K.

FIGURE 3.5 "Geographical Puzzle," *Harper's Young People*, November 26, 1889 (courtesy of Library of Congress).

State and *falls in Africa*, were sent on a myth-bound, originally Native American practice, namely a buffalo hunt. North American buffalo being nearly extinct by 1890, Dare and Frank K. relocated the fictional brothers to Africa. Their story confides in their readers that after the brothers failed to find any buffalo, they instead encountered a lion, which *a lake in New York State* managed to slaughter. On his way home, *a lake in New York State* got a Cashmere goat and a Shetland pony for his sisters, a choice that underscores the importance attached to the idea of "home": pets, siblings, and family.[28] Unfolding in a historically telling sequence of events (the western frontier closes and the buffalo go extinct, after which Americans go abroad, hunt African lions, and bring back to the United States souvenirs from across the world), the story not only marks children safe as they go on adventures away from home but also remaps the U.S. empire away from its continental borders and places it at the beginning of a new chapter in its rise to global empire following the closing of the western frontier.

On the face of it, it might seem as if it was their active imagination with regard to the teachings of home geography that made it possible for Frank and Dare to name their story's protagonists after natural features in locales as near as New England and as far as Africa. And yet I would like to insist here that much more active thinking and maneuvering went into their choice of names. It is hard to believe that Frank and Dare decided on their protagonists' origins as African and American as a conscious hint on the wished-for blurring of racial boundaries in the postemancipation United States. Even so, as the answer to this puzzle reveals, Frank and Dare's choice of names has a complex backstory to it. This puzzle was possible to compose (and solve) because children knew about and consumed popular sources of geography other than their primers and readers. The brothers in the puzzle are thus named George and Stanley—as in Stanley Falls, the Congo. The Stanley Falls were named after Henry Morton Stanley (1841–1904), a Welsh-born American journalist, sailor, and explorer famous for his explorations in Zanzibar and along the Congo River. His many expeditions in the heart of Africa had brought Stanley immense wealth and fame, especially when in 1871 his secret *New York Herald* commission in Africa to look for the missing world-renowned Scottish explorer and missionary David Livingstone (1813–73)—who had been commissioned by the British government in 1866 to find the source of the Nile—was successful. With his extensive network of friends and agents among Arab slaveholders and African chieftains, Stanley's expeditions were later commissioned by the Belgian king Leopold II and were, more than ever, at the service of European colonialism on the "dark continent."[29] *The Boy Travellers on the Congo*, written by the American explorer-turned-journalist Thomas Wallace Knox (1835–96) and published by Harper and Brothers in the series' 1887 volume, had made Stanley a household name across the United States.[30] By the time Knox's book was published, Stanley was "famous throughout the civilized world for the great work he has performed in exploring the African continent and opening it to commerce and Christianizing influence."[31] In fact, Frank, Dare, and other children in their generation were able to imagine two American brothers in one family whose names were taken from places as far from each other as New England and Africa and to find the right answers to such puzzles partly because they had been exposed to exceedingly popular sources of world geography such as the Boy Travellers series.

In a broader historical perspective, Knox's praise for Stanley involved crude applause for a practice that had, across several centuries already, made various parts of Africa "familiar" to European eyes and ears—quicker to spot on table globes and world maps and easier to pronounce and to spell. Indeed, when Stanley explored, located, and named as the Stanley Falls a group of seven cataracts in central Africa (currently known as Boyoma Falls) near where the Lualaba River meets the Congo River, he took part in this long tradition of renaming native spaces into colonial places.[32] In the colonization process, renaming enabled the colonizer to establish an unequal power relation to colonized spaces, both in the colony and in the imagination of the untraveled average citizens of the metropole. This involved robbing colonized people of a sense of familiarity, safety, and belonging to and owning space—including their very own bodies. Therefore, the practice of renaming colonized spaces was both geo- and biopolitical. Moreover, it was no less significant and harrowing than acts of colonial cartography, including "cleansing" regions formerly unknown to the colonizers as uninhabited through labeling them terra incognita, or mastering the body of the colonized by forcing upon them Western customs, costumes, and languages. Indeed, christening spaces with Western names and Christianizing their so-called heathen populations were hand-in-glove projects of colonization—part and parcel of what Mary Louise Pratt refers to as an imposed, colonial "planetary consciousness."[33]

It is inaccurate, of course, to reduce the complex, violent, and violating tangle of colonizing measures of the European and later American empires to mere acts of renaming places and peoples. Colonialism, even if examined exclusively in terms of cartography, entailed redrawing political borders and erecting new spaces on the ruins of the old, imposing development plans on the landscape—such as sketching resettlement projects or designing canals and roads before such plans became physical realities—and of course shifting the contours of political allegiances by changing the color codes or placing arbitrary lines on maps and globes. However, a considerable portion of colonization's ideological work has always been done through renaming. There is enormous political sense to extract from acts of renaming, including historical and legal distortion, alienation and resentment, adjustment, and resistance. As chief agents of renaming the colonized, eighteenth-, nineteenth-, and twentieth-century Christian missionaries renamed members of the communities they wished to "civilize" not only to, quite pragmatically, make the daily interactions with and the task of writing reports about those individuals easier, but also to underscore the success of the white man's civilizing mission. Such a seemingly simple practice would also change the colonized populations' relationship to their individual, un-Christian pasts, exert power over their identity, and force upon them a new, alien identity. Generations of enslaved people and members of various Native American tribes in colonial America and later the United States were also renamed by white masters or teachers, not only to be reduced to the level of the white slave-owner's property but also as an involuntary rite of passage into slavery and submission—an act that converted even their given names into sites of racial struggle and resistance against their white "masters" and overseers.[34]

Back in the metropole, and thanks to a world that was both familiar and accessible, adults could map themselves as part of a homogeneous colonial network that, although largely faraway, was meticulously mapped, pinned in place on table globes, and rendered legible. As Frank and Dare's story puzzle illustrates, children emulated this mapping practice, as a consequence of which they could embrace adult colonizers' mapping habits by playfully devising and solving geographical puzzles. To invoke Fredric Jameson's "cognitive mapping" as a collective practice, at the peak of the age of empire, both mapping and renaming colonized spaces could be understood as aesthetic, surgical measures taken in the metropole to allay the confusion caused by the buildup of innumerable geographic particulars and cartographic minutiae of a by-and-large colonized world. Making sense of the colonial hyperspace required cognitive mapping of the world—not only as *reflective* of subjective spatial practices but, more importantly, as *performative* of collective political moves. After all, cognitive mapping is a practice that conflates geographical facts and collective spatial sensibilities and predilections with regimes of taste and desire scripted as individual experience.

Beyond doubt, colonial names complicated the relationship of a child called Alphonse, Stanley, or Florence with world geography at the same time that they cemented their sense of distance between these places and their home. With children recognizing homonymy between colonial names and their own, their friends', even their pets' and dolls' given names, renaming could, in the representational sense that David Harvey understood appropriation of space (see Figure 1.1), produce intimate geographies of the colonized world among children of the metropole.[35] This was a subjective, playful, and possessive relationship that helped the construction of more personalized and accessible cognitive maps among the youth. From the subjective perspective of young American puzzle composers, American children got to know their given names first as arbitrary yet intimately passive features of their identities, assigned to them by their parents at birth, and only later as the toponyms they found on the world map or in the home atlas.

Mapping projects involve various levels of abstraction in order to simplify and bring order and stability to an otherwise shifting, fragile, and incomprehensible whole. By the same token, the names that appeared in these puzzles—though primarily geographical—were in fact (re)abstracted from real territories, forced to assume British, German, or Spanish proper names, which children could then convert into tangible objects in their lives: pets, playmates, food.[36] These regions' dense histories of (in-)dependence, the hostility or alliances among their inhabitants, and the colonial contexts in which new names such as Elizabeth or Alphonse were forced on them had no space on European and American cartographic narratives. In fact, the story puzzles' patchy choice of names from different localities across the globe, noted or obscure, American or Asian, make it abundantly clear that such choices were made not to reflect a knowledge of local historical and geopolitical intricacies but in accordance with their composers' random fingertip journeys across a globe with which they were more or less familiar—a world in which they took an interest, even if only minimally. The exchanges between these randomly picked localities on the world map were not necessarily a given at the turn of the century, or at least not directly apparent and relevant to the intended purpose of these puzzles. It was, in fact, children who plucked geographical place names and natural features from atlases, maps, or globes and bound them into narrative wholes wherein distant places, mountain peaks, and riverbeds on other continents would make it possible for the stories they wrote to take place in an American setting and for their authors to spend an uneventful, apolitical afternoon in the U.S. countryside—by way of world geography.

Partly internal and inaccessible as it is, such a process of abstraction invokes Jean Piaget's classic, albeit often criticized, observation that children's desire to name a place equates the desire to know it. This desire emanates from a myriad of conscious and unconscious geopolitical associations that children make simply by recalling a place's name.[37] How Alphonse *could* show up in the unsigned short-story puzzle analyzed at the beginning of this chapter as the indirect result of a narcissistic naming act and a colonizing gesture by a French commander has a lot to do with the co-optation of place names and geographic knowledge. Beyond doubt, in the power imbalance between the colonizer and the colonized, between the "namer" and the "named," there always lies a multitude of superimposing yet not necessarily corresponding and connected worlds that serve as reference points for human action and knowledge. Names demand consensus and make navigation of multilayered spaces of interaction possible, joining the colonizers' triumphant exploitations with the colonized populations' traumatic experiences.[38] Indeed, to "know" a place name—to hear one's own name in it, or to think it familiar enough to be used as a puzzle clue—inscribed American children in the long practice of colonial renaming as an adult, violent practice that involved spatial decision-making on four of the levels (accessibility, appropriation, domination, and production of space) that Harvey identifies in his grid (see figure 1.1).

Alphabetical Puzzles and Synecdoche

Another popular form of geographical puzzles created by children for *Harper's Young People* was the alphabetical puzzle. In composing alphabetical puzzles, children allow us a partial and parsed access to their intimate understandings of world geography not through the sound of names but rather by establishing synecdochical relationships between individual letters of the alphabet—the very fundaments of child literacy—and the place names they spelled. In my reading, the result of these predominantly unrecorded and fleeting but significant moments of deciding what place names A or E or Y spelled was the outcome of distinctive patterns by which children tended to process geographic information both about the United States as their home and about the world as its distant parallel. But first we need to interrogate the ways that letters of the alphabet as empty signifiers and writing have been recruited in the West to signify complexly geopolitical, race- and class-conscious practices and policies.

Over the past two to three decades, letters of the alphabet have been studied in a number of historical contexts with regard to their capacities both as the building blocks of literacy and as tools at the service of cartographic practices. As carefully illustrated by Patricia Crain, letters of the alphabet have been assigned this significant dual role because a wide range of social, political, and cultural forces "help to shape the alphabet's discursive space."[39] Letters have historically been perfected into culturally agreed-upon written signs that represent utterable sounds in a great majority of linguistic communities across the globe. Likewise, they have made the practice of recording ideas and of transmitting meanings within a community and across generations consistent, accessible, and durable. Those individuals and institutions that could make use of such arbitrary signs—a powerful minority that, over a considerably large proportion of human history, tended to grant access to the privileges of literacy to only a select few—were able to assign new meanings to strings of, or even to individual, letters of the alphabet in order, for example, to mark movements, tag singularities, erect categories of difference, or exclude actors and entities. Kathleen Biddick's close examination of several medieval Christian maps and travel books, for instance, shows how systematically and effectively their random use of Hebrew letters transmuted Jews into Others of Christendom and rendered them temporally absent as Christians' contemporaries.[40] In "The ABC of Ptolemy," Biddick analyzes Christian texts and maps in which otherness has been coded through the meticulous placement of Hebrew letters in the margins of maps. These medieval maps, Biddick asserts, entailed complex cartographic practices—interchangeably employing the "alphabet as territory" and "cartography as alphabet"—practices through which Christian communities managed to visually other, collectively earmark, and ultimately physically expel Jewish people from Christian territories, starting with doodling Christian maps' margins with spatially vagrant letters of the alphabet of the "undesirable."[41]

Accordingly, letters of the alphabet have a long history of cartographic functionality in the West, at once expressing xenophobia and delineating territories, since at least the Middle Ages.[42] Sam Loyd's puzzle *Get Off the Earth*, as we have seen at the beginning of this chapter, is an outstanding example of this practice in modern times (see figure 3.1). It highlights how letters of the alphabet could be applied to visually sign and discursively legitimize the rejection and expulsion of the "unwanted, foreign element" from U.S. society (*and* the world). In *Get Off the Earth*, the letters carelessly drawn on either side of the puzzle's globe were not real Hanzi characters but only a rough, offhand imitation thereof.[43] Lazy imitations, they were meant to accentuate the otherness of Chinese immigrants via the sloppy replacement of Hanzi characters with a sketchy, erroneous set of doodles. This careless squiggling further marks a general lack of interest in and knowledge about certain groups of outsiders as white Americans' ultimate Other.[44] By placing illegible approximations of Hanzi characters around the image of a globe, Loyd projected "the Sino-American question"—specifically, the opposition to Chinese immigration to the United States, which served as a presidential campaign motif—onto a global scale, attaching Americans' nativist wish that Chinese people simply disappear from the United States to a seemingly universal desire to be rid of them, as if that were a desire shared by all humankind. Ultimately, as I argued earlier, the contrast cannot be starker between the violence exercised on the warriors' bodies as they are cut into halves thanks to the puzzle's mechanical precision and the indifference the puzzle demonstrates in placing a handful of make-believe Hanzi characters in the puzzle's margins.

As Crain's in-depth study of the alphabet as the prime medium of child socialization reveals, beginning in the seventeenth century, children of British North America learned the alphabet from *The New England Primer* and other ABC books in which strict regimens of moralizing human action (A for the sinful Adam) as well as memorizing gender-specific roles (S for Sewing) and adult-appropriate class-based occupations (C for Carpenter or K for King) were spun around individual letters as examples.[45] In the mid-eighteenth century, and as children began to be addressed as a new target of moral and, later, patriotic messages, the alphabet "was dressed up and decked out, animated, ornamented, narrated, and consumed."[46] In other words, it was transformed into a vehicle for transmitting social as well as ideological content. By and by, Crain concludes, the alphabet came to function as carriers of material meaning and to stand for external objects in the world and give order to them. In the nineteenth century, for instance, "both representative of and represented by the world, the new alphabet [became] a kind of world in itself."[47] By the end of the same century, the alphabet arrays that appeared in common children's primers had ceased to signify morally charged words or to list deadly sins and their subsequent punishments.[48] Children who memorized the letters of the alphabet in their given order were now taught to name material objects like household appliances, food, human emotions, or socially acceptable occupations, as if each letter had a simultaneously part-to-whole and one-to-many relationship to the seemingly endless list of mono- and polysyllabic words for which it could potentially stand.[49] Depending on the synecdochical nature of the letters of the alphabet, endless lists of words beginning with the same letter of the alphabet began to fill young children's horn books and ABC books.

Back to juvenile periodicals, it is evident that during the final ten years of its publication, almost every issue of *Harper's Young People* included an alphabet-based geographical puzzle. It was the most common type of geographical puzzle composed by children, perhaps because it did not require any additional writing skills, as were needed in the case of story puzzles. In essence, alphabetical puzzles were lists of simple geographical statements such as "A is a city/a mountain/a country/a region in/near some better-known place," where A was usually the toponym belonging to—hence smaller than and politically dependent on—the better known and more meticulously mapped B. The alphabet, in these puzzles, was at the service of a new understanding and function of world geography that American children studied, read about, and played with. This world consisted of an assortment of homes that were indexed by the alphabet array. Synecdochically, then, the new alphabet had come to stand for the world itself.

In accordance with Piaget's findings, when a child asked in these alphabetical puzzles for a toponym starting with F or O or Z (a toponym understood to be part of a larger spatial unit), this did not necessarily mean that the child, or other children who would then solve the puzzle, had developed a clear, linearly nested understanding of the geographical scales that linked smaller geographical entities to the countries, continents, or bodies of water in which they are located.[50] Piaget's founding studies indicate that younger children have difficulty understanding and accurately visualizing what a statement such as "Geneva is in Switzerland" means. It was not a given for Swiss children in Piaget's studies to be both from Geneva *and* from Switzerland. For instance, when asked to draw Geneva and Switzerland in relation to each other, children seven years of age or younger drew two circles—sometimes one bigger for Switzerland—but side by side, not one inside the other.[51] This pattern changes, the studies confirmed, as children grow older. Perhaps equipped with a spatial knowledge that they acquire through schooling, encounters with maps, or family trips, children between the ages of seven and eleven begin to understand that Switzerland is larger than and envelopes Geneva. Likewise, children eleven years of age or older, the study concluded, have a clearer understanding of their nationality, can locate Geneva as spatially smaller and nested in a whole called Switzerland, and have a more or less solid sense of belonging to an unseen community of Swiss people.[52] In connecting Piaget's findings to my reading of the formulaic statements in the alphabetical puzzles, I conclude here that under the influence of family, playmates, schoolmates, and society in general, children develop ever more complex, not necessarily linear, conceptions of their home and of its links to other homes, their town and other towns, their country and other countries, the continent where their country is located and other continents, and, ultimately, the world.

A striking example of this pattern is on display in an alphabetical puzzle by Sphinx, printed in the May 20, 1890, issue of *Harper's Young People* (figure 3.6). In this puzzle, twenty-six letters of the alphabet stand for twenty-six cities in twenty-six different countries. Sphinx's puzzle consists of an orderly list of questions, each about examples of the same spatial unit—that is, the city—within a common political denominator: the nation-state. The puzzle's solutions are scattered across the globe: A stands for a city in Australia, I for a city in Persia, L for a city in Portugal, and Z for a city in Switzerland. Directing puzzle solvers to look for the "right" toponyms all over the globe, this predominantly Eurocentric puzzle decides that almost three-fourths of the letters of the alphabet should stand for cities in Europe, from Norway to Italy, and from England to Greece. The eight remaining letters represent cities in Australia, India, Persia, Egypt, Zanzibar, China, Algeria, and Russia. In terms of points of reference on the map of the world, this is one of the simplest and most structured of all the alphabetical puzzles printed in *Harper's Young People* during the 1890s. Notably, no mention is made in this puzzle of any American cities, large or small. Why are American cities absent from Sphinx's puzzle? One immediate response we might have to this question is that the puzzle composer intended to make the puzzle less commonplace and obvious.

No. 4.
ALPHABETICAL CITIES.

A is a city in Australia.
B is a city in Norway.
C is a city in Ireland.
D is a city in India.
E is a city in Scotland.
F is a city in Italy.
G is a city in Belgium.
H is a city in England.
I is a city in Persia.
J is a city in Roumania.
K is a city in Egypt.
L is a city in Portugal.
M is a city in Germany.
N is a city in France.
O is a city in Denmark.
P is a city in Greece.
Q is a city in Zanzibar.
R is a city in Russia.
S is a city in China.
T is a city in Austria.
U is a city in Holland.
V is a city in Bulgaria.
W is a city in Algeria.
X is a city in Spain.
Y is a city in Japan.
Z is a city in Switzerland.

SPHINX.

FIGURE 3.6 "Alphabetical Cities," *Harper's Young People*, May 20, 1890 (courtesy of Library of Congress).

Similarly, Sphinx might have wished to showcase their knowledge of world geography beyond U.S. borders and to make the act of solving the puzzle more adventurous. Another reasonable explanation for the exclusion of the United States in this puzzle may be that it reflects an undisputed, indeed far from puzzling, societal consensus as to where on the map of the world the United States stood and which prominent place names dotted it. As mirrored in this puzzle's choice of questions, even at the turn of the twentieth century, the most common view of the nation was that—with its roots mainly in Europe—the United States was still geographically disparate and politically poles apart from the rest of the world and yet had a clear impression of where in the world its commercial and political interests lay. Following school geography primers, at least in the case of this puzzle, the composer may have chosen place names as if planet Earth consisted of two worlds side by side: one

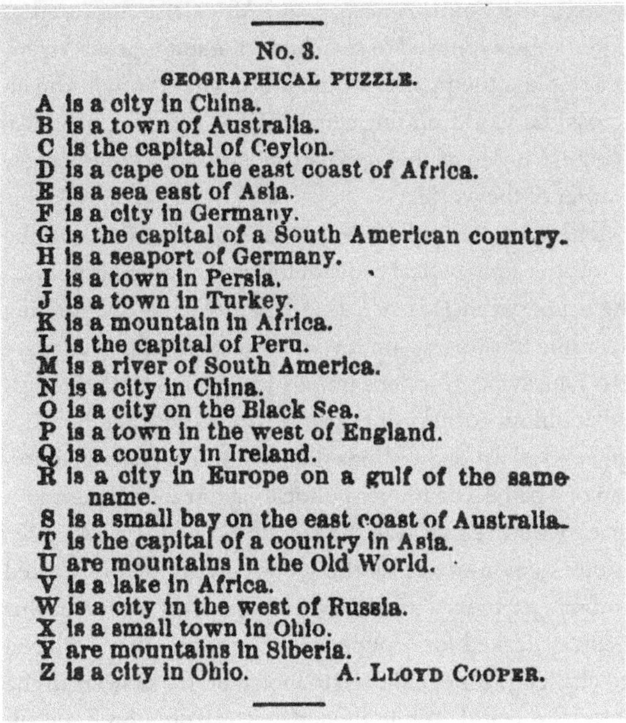

No. 3.
GEOGRAPHICAL PUZZLE.

A is a city in China.
B is a town of Australia.
C is the capital of Ceylon.
D is a cape on the east coast of Africa.
E is a sea east of Asia.
F is a city in Germany.
G is the capital of a South American country.
H is a seaport of Germany.
I is a town in Persia.
J is a town in Turkey.
K is a mountain in Africa.
L is the capital of Peru.
M is a river of South America.
N is a city in China.
O is a city on the Black Sea.
P is a town in the west of England.
Q is a county in Ireland.
R is a city in Europe on a gulf of the same name.
S is a small bay on the east coast of Australia.
T is the capital of a country in Asia.
U are mountains in the Old World.
V is a lake in Africa.
W is a city in the west of Russia.
X is a small town in Ohio.
Y are mountains in Siberia.
Z is a city in Ohio.   A. LLOYD COOPER.

FIGURE 3.7 "Geographical Puzzle," *Harper's Young People*, January 7, 1890 (courtesy of Library of Congress).

that included Australia, Asia, Africa, and Europe, and the other that included the Americas.[53] Centering the puzzle's clues mainly on Europe and completely avoiding any mention of the United States make it clear that the puzzle was not made by casually paging through the list of place names at the end of the puzzle maker's geography book or home atlas—lists that would also have included a number of American toponyms—but by a more watchful process of selection, one that subtly reflected American's geopolitical priorities beyond the borders of the nation.[54]

Unlike Sphinx's puzzle, not all alphabetical puzzles were consistent in their points of reference. *Harper's Young People*'s January 7, 1890, issue included a geographical puzzle containing twenty-six puzzle statements that asked for the names of a host of cities, regions, and natural features as parts of a variety of larger geographical units, from cities to bays and lakes, in the United States and elsewhere (figure 3.7). A. Lloyd Cooper, the composer, referred in this puzzle to Europe more than to any other place, including the United States. The puzzle also referred to China, Russia, Australia, Persia, Asia, and Ohio, twice each. Turkey, Germany, South America, and Africa also appeared as places in which children were supposed to search for mountain peaks, rivers, or cities in order to complete the puzzle. While journeying through the alphabet array and across the world on fingertip, this and other alphabetical puzzles posited the empty order of the alphabet in contrast to the densely packed geopolitical order of the world.

In other words, letters of the alphabet in these puzzles indexed the world. At the same time, the alphabetical order of these puzzles made geography secondary to the primary function of listing, ordering, and categorizing worldly objects—including territories—on the world map. Looking for place names from Amoy to Zanesville, Cooper's puzzle, for example, asked questions about seemingly random corners of the world, more with the purpose of completing its alphabetical list of questions than to unveil a concrete internalized cognitive map or a pattern of any geopolitical significance (figure 3.8). No single continent or world region was the sole focus of the puzzle. No single European empire's colonies across the globe, or any of the United States' potential economic partners—that is, the global market for its surplus products—were specifically asked for. Cooper's list's purpose does not even seem to have been to challenge other children to look hard for answers in the least known corners of the world. While the letter X usually posed a challenge for young puzzle makers—a challenge solved both here and in another puzzle from January 28 of the same year by making it stand for a town in Ohio—other letters stood for random places and were, more often than not, easy to figure out. In other words, the alphabetical array in this puzzle seems to have overshadowed the cartographic functions it might also have had, meaning it was the alphabet that dictated where on the map of the world the child should look for the right place name, rather than the other way around. Of course, not just *any* place of interest on the world map would do. The encyclopedic listing feature of the alphabet array, as the latest and most common feature of ABC books since early 1800s, seems to have overridden geography in the imagination of many a young American child, as is evident in this and several other *Harper's Young People* puzzles in the early years of the 1890s.[55]

In effect, the encyclopedic logic of these puzzles seems to have been at once bolstered and overruled by the idiosyncrasies that each puzzle maker displayed in selecting the set of geographical reference points that the same letters of the alphabet could stand for in different puzzles. After all, each set of geographical reference points would be charged with geographical and geopolitical significance that were unique to that very puzzle according to its composer's choices. This in turn mirrored, at least partially and perhaps arbi-

> No. 3.—Amoy. Brisbane. Colombo. Delgado. Eastern. Frankfort. Georgetown. Hamburg. Ispahan. Jiddah. Killmanjaro. Lima. Madeira. Nanking. Odessa. Preston. Queen's. Riga. Shoalwater. Teheran. Ural. Victoria. Warsaw. Xenia. Yalbonoi. Zanesville.

FIGURE 3.8 Answer to "Geographical Puzzle," *Harper's Young People*, January 28, 1890 (courtesy Library of Congress).

trarily, the composer's cognitive map of the world at the time of composing the puzzle as well as the individual strategies they used to make a puzzle stand out as more challenging or as inclusive of more singular and unprecedented place names. To take the example of A, for instance: while for Cooper A best exemplified Amoy (a city in imperial China), Laura P. Smith (July 1, 1890), Mattie Steiner (December 23, 1890), and Estelle Lidenburg (February 10, 1891) all thought of A as the river Amazon. Ella F. (July 22, 1890) and Jeanne Lancaster (August 26, 1890) started their alphabetical puzzles by referring to Austria, while A stood for the Adirondack Mountains in B. R. E.'s puzzle (September 16, 1890). Moreover, Ava, a city in what was then Burma, was what the letter A represented in Elsey Swasey's puzzle (November 25, 1890). Examining the diversity of the names for which a single letter stood in this sample suggests that a rather subtle pattern existed behind the seemingly random choices different children had made. While it was not necessarily an insurmountable puzzle-solving challenge to find closer-to-home place names beginning with A, these puzzles suggest that their composers followed a pattern in asking for names in those parts of the world with which they were more familiar—that is, Europe and the Americas.

In the same vein, with regard to the Western Hemisphere, for instance, children typically asked for the most famous places and the most prominent natural features—names they might have come across, collectively or individually, in daily conversations; on family trips; or in geography books, travel guides, or dime novels. On the other hand, the further east and southeast they looked on the Mercator projection, the less repetitive and predictable their references became. Indeed, when asking for place names that a single letter stood for in faraway parts of Asia, Africa, or Australia, children's dependence on the world map (with their distortion of size and scale) and their lack of hands-on knowledge about and interest in those continents would transform them into explorers without itineraries, finger-tip travelers who stumbled upon place names by accident. Depending on where children looked for geographical names, the answers to their puzzles either exhibited a pattern of signposting sites considered geopolitically important to the United States or diverged into scattered reference points on the map of the world. In sum, in composing these puzzles and the fictive globe-trotting adventures that they entailed, American children were at times selectively encyclopedic and at others adventurously indiscriminate.

Scale and the infrastructural nature of geographic reference points signify another set of patterns, too; on the one hand, for example, Cooper's puzzle zoomed in to ask for the name of a county in Ireland, and their X and Z stood, respectively, for a small town and a city in Ohio (see figure 3.7). In these cases, limited with choices for X and Z, young Cooper referred to modern spatial units of division, which denoted administrative hierarchy, judicial order, and urban industrial discipline—all characteristic of an established Western civilization, with its rush toward urbanization and away from the dictates of nature and premodern communal life as deemed to be the dominant order of life in the non-Western world. On the other hand, in the same puzzle, E was a sea east of Asia, and K and V, respectively, a mountain and a lake in Africa. As an American growing up in the 1890s, Cooper's choices were in line with common perceptions of the time regarding the implications of the closing of the western frontier as opening the doors to a world in which Americans defined more or less in terms of the urban West and the nonurban rest—a theme that repeated itself throughout other puzzles, too. Older civilizations, such as the Chinese, Indian, and Persian, though all in the East, were mentioned in this puzzle in reference to their old, major cities, as was the case with European cities such as Hamburg, Warsaw, and Frankfurt.[56] But Latin America and especially Africa were referred to as undivided wholes whose allegedly nonurban, uninhabited, politically undifferentiated, and open-to-colonization landscape were peppered with mountain peaks, rivers, and lakes named tellingly after European explorers. Unlike in their questions about the United States and Ireland, the young puzzle maker cast a cursory, withdrawn yet enterprising look at Africa from a distance at which only a mountain such as Kilimanjaro (misspelled by Cooper) and a lake such as Victoria were visible.

A comparable puzzle by S. Wolleson, printed as "Geographical Puzzle No. 6" in the January 28, 1890, issue, asked for fifteen cities, one town, two countries, and eight natural features. In this puzzle, X again stood for a city in Ohio. However, Z did not represent the Ohioan city of Zanesville anymore; rather, it took the child puzzle solver to Africa and asked for a country off the continent's east coast: Zanzibar. While choices for geographical names that start with Z are limited, it could be that Zanzibar's "exotic" name had earned it an entry in S. Wolleson's list.[57] However, Zanzibar most likely appeared in this puzzle as a result of the craze over further colonization of central Africa, which, at least for American children growing up during the last two decades of the century, had an American face to it in the figure of Stanley. After all, following a long clash of interests between British and German colonial rule over Zanzibar in the midst of the so-called Scramble for Africa, the Helgoland-Sansibar-Vertrag, also known as the Anglo-German Agreement of 1890, was signed between the two European powers, ensuring the end of German interference with British rule over Zanzibar in exchange for Germany's naval control over the Kiel Canal in the North Sea.[58]

S. Wolleson's puzzle statements reveal a strict West–East division not too different in its basics from the one in Cooper's puzzle. Seventeen of the twenty-six references in S. Wolleson's Eurocentric puzzle pointed to European cities, seas, or rivers. The easternmost place from the United States was A, "a city in Turkey," while the closest place to S. Wolleson's home was Y, "a town in New York," the only letter of the alphabet that stood for a place in the United States. Thus, it seems that the puzzle maker was most interested in Europe and Latin America and showed little interest in Asia and Australia. It is also likely that S. Wolleson made the puzzle statements not by checking random names on a globe but rather by taking them from a school geography primer or home atlas. If following a book or a curriculum that progressed from West to East, they might have learned about the Americas, Europe, and Africa but not yet reached lessons on Asia and Australia in geography class.

It is noteworthy that, here again, as with the previous puzzle and more than a few of the dissected maps studied in chapter 2, references to the United States were less frequent and at the same time were made on a much more meticulous scale than in the case of references to other parts of the world. Laura P. Smith's and Ella F.'s puzzles, printed, respectively, in the July 1 and July 22, 1890, issues of *Harper's Young People*, reveal comparable patterns and scalar convulsions. Laura P. Smith's and Ella F.'s puzzles referred to different parts of a large, seemingly undivided geographical unit as Africa, showed an apparent preference for Europe, asked for cities in different European countries, and mentioned Latin America and Africa only in relation to their fascinating landscapes. As a rule, Laura P. Smith zoomed in to ask for the names of cities and rivers in individual European countries, while her questions about other parts of the world were usually in reference to a city, a cape, or an entire region located within the whole width of a continent such as Africa. In Ella F.'s puzzle, the scale was far more zoomed in when she asked for cities in different American states. Such a stark shift between the scales used in reference to localities in the United States versus places outside its borders repeats itself in

> No. 6.
> GEOGRAPHICAL PUZZLE.
> I am composed of 15 letters.
> My first is a gulf west of Russia.
> My second is a cape off Newfoundland.
> My third is a river in South America.
> My fourth is a town in Rhode Island.
> My fifth is a cape off Ireland.
> My sixth is a country in Africa.
> My seventh is an island near Italy.
> My eighth is a bay in Canada.
> My ninth is a bay on the coast of France.
> My tenth is a great republic.
> My eleventh is a river in Germany.
> My twelfth is a sea east of England.
> My thirteenth is a channel between England and France.
> My fourteenth is an island south of Australia.
> My fifteenth is a town in Japan.
> The initials spell the name of a popular living authoress. MARIE BUCHANAN.

FIGURE 3.9 "Geographical Puzzle," *Harper's Young People*, April 8, 1890 (courtesy of Library of Congress).

puzzles printed later in the decade, such as Jeanne Lancaster's, in which W stood for mountains in Utah and J represented a city in Virginia, and B. R. E.'s puzzle, in which several letters of the alphabet stood for mountains, cities, rivers, and lakes across the eastern United States. Considering the various location distribution patterns at work in these puzzles, to draw a world map based on individual puzzles is at once tempting *and* unfeasible exactly because the composers invoked incongruent scales in thinking and asking about places at home and abroad. In effect, children seem to have fluctuated between locating A as a place name within the entire expanse of China and Z as an industrial town in the State of Ohio.[59]

Next to alphabetical puzzles, acrostic and double acrostic puzzles present other patterns in addition to the East–West patterns of division and progression previously analyzed. One acrostic puzzle, composed by Marie Buchanan and printed in the April 8, 1890, issue of *Harper's Young People*, consisted of fifteen puzzle statements (figure 3.9). Marie Buchanan's puzzle did not ask questions for every letter in the English alphabet. The answers' initials, this geographical acrostic maintained, would spell the name of a contemporary female author—that is, Frances H. Burnett (1849–1924). Marie's puzzle is more revealing than the puzzles analyzed so far, because with its selective listing of the letters of the alphabet, the acrostic is a completely different form of puzzle. Marie had the option of choosing examples for the fifteen letters of the alphabet she had in mind for the final puzzle question from a wide range of places or geographical features. Furthermore, she asked more than one question about the same letter of the alphabet, and she did not have to worry about tricky letters such as X and Z. Thus, combining her literary and geographical interests and knowledge, her puzzle could overcome the dictates of the alphabetical-geographical correspondences that the previously examined alphabetical puzzles had to sort out.

All letters except the first E and the U in Frances H. Burnett's name referred to a smaller region or geographical feature, such as a city, an island, or a sea, in or near a larger geographical unit, namely, a state, a country, or a continent. In the case of the English Channel, for instance, the second E connected Britain to the Continent. Marie Buchanan's puzzle was the first puzzle *Harper's Young People* printed in the 1890s in which the United States was mentioned *not* to ask for a city in one of its states or for a river flowing through the Midwest. Rather, the clue "my tenth is a great republic" looked for a country that was neither marked by Marie on the map of the world nor cartographically adulterated by an excess of toponyms and border or railway lines. This "great republic" that provided the U in Burnett did not refer to a lake near a city or a country in a continent. Rather, it pointed to a transgeographical entity that was characterized by its exemplary sociopolitical system. The outright unspatiality of the United States in this geographical acrostic puzzle points to the predominantly exceptionalist discourses that informed Marie Buchanan's perception of the United States. Instead of "my tenth is a country in North America"—a statement ostensibly comparable to the rest of the statements in this and other alphabetical puzzles—Marie in fact presented her country as standing alone, in and of itself. This great republic, this exceptional home, was not nested in any larger geographical unit. It did not neighbor any other region or natural feature, nor was it in fact geographical to begin with. It was Marie's (and Francis Burnett's adopted) home; a space that, while purposefully written off the map of the world, pointed to a specific collectivity of people, called Americans, and their sociopolitical achievement, namely their republic. As if separate from the unchangeable dictates—or given blessings—of nature, Marie's U stood above geography.

A prolific writer, Frances Hodgson Burnett (1849–1924) was a British immigrant to the United States. When she was sixteen, her family settled in Tennessee, where she dedicated herself to writing works of fiction for children and adults alike. In the 1890s, Burnett returned to London, but she made frequent trips to the United States until her death.[60] Though she wrote primarily for an American audience, Burnett's appearance as the final answer to a geographical puzzle printed in a best-selling American juvenile weekly reminds us of a fact that is central to a great number of these child-crafted puzzles: that the words that the letters of the alphabet stood for did not simply bind children to the typically heavily sheltered spaces of childhood within the borders of the national, such as home, orphanage, kindergarten, school, or hometown; rather, they sent them on journeys across the globe and in search of what was far off, unknown, and colonized.

Marie Buchanan's puzzle was composed before Burnett left the United States to settle in London.[61] At this time, she was considered a quintessentially American author by her young American fans. Consequently, Marie Buchanan decided, perhaps out of patriotism, that the letter U in her beloved author's name had to stand for the United States, her adopted home. Burnett in this puzzle represents the many unnamed Anglo-Saxon people who were, white Americans believed, behind the project of maintaining the United States as a great republic. The contrast cannot be more telling when we juxtapose the letter's attempt to celebrate Burnett as the author of their beloved *Little Lord Fauntleroy* with the commonly known facts about the intensely troubling acts of racial violence she had done to the black doll she owned as a child when reenacting the whipping scenes from *Uncle Tom's Cabin*.[62] Back with the imagery implied by home geography, the republic that Marie Buchanan wove Burnett's name into was the very republic that had for long sanctioned systematic violence to its Black population—the Black Topsy whipped, raped, and then tucked away into the most intimate of domestic spaces, that is, under the white Turvy's skirt.[63]

On a more directly geographical level, Marie Buchanan's attempt at collecting the letters of Burnett's name by referring to places as scattered across the surface of the earth as Finland, Egypt, the Hudson Bay, the English Channel, and Tokyo would give Burnett's representative name a transgeographical aspect, as if her name were a poetic synonym for "global"—a homonymously familiar geographical scale clad-as-proper-name that took puzzle solvers to different corners of the world in search of the right letters of the alphabet, which, synecdochally, stood not only for physical places but also for a great republic built by the choicest immigrants from those very corners. Presented as the product of a liberal thought that was fulfilled through a collective practice and sustained by an assortment of migrants from near and far corners of the world, Marie Buchanan's United States divorced itself from colonial settlement, warfare, slavery, and aggressive continental and overseas expansion. In a sense, in this puzzle, the "republic of alphabet" coincided with the U.S. republic, and together they facilitated bringing up globally cognizant, patriotic

> No. 6.
> PIED CITIES.
> 1. Dllceevna. 2. Obflafn. 3. Yknoewr. 4. Bspetegtrusr. 5. Ghgonkno. 6. Ecnstoiapotnln. 7. Apihhipllaed. 8. Oollpirve. 9. Mbeitalro. 10. Sowelnraen.
>   RALSTON SEYMOUR.

FIGURE 3.10 "Pied Cities," *Harper's Young People*, March 4, 1890 (courtesy of Library of Congress).

American children, a role Marie Buchanan enacted in selecting her clues and answers.[64]

Letters of the alphabet made further appearances in disorderly and hidden ways in other types of puzzles. In the case of pied puzzles, the letters of the alphabet that made up the names of cities, countries, or natural features were scrambled within an individual word. Examples such as "Ghgonkno," "Apihhipllaed," and "Mbeitalro" would need to be rearranged to spell, respectively, Hong Kong, Philadelphia, and Baltimore (figure 3.10).[65] As an example, Ralston Seymour's pied puzzle asked for the names of ten urban centers up and down the East Coast of the United States as well as in Russia, the Ottoman Empire, the British Empire, and Britain. No explanation preceded or followed the scrambled words. Six out of ten of these place names were those of prosperous U.S. urban centers, underlining an awareness of the United States' rising position as a modern commercial power on the part of the puzzle's child composer. However, references to cities/regions such as Hong Kong, a contested British possession since the end of the First Opium War in 1842, and to Constantinople, the stronghold of the Ottoman Empire's political and economic relations with Europe and the United States in the form of banking, infrastructure building, tourism, and missionary activities, also pointed to the significance of non-American, imperial urban centers for the puzzle's young American composer. During the course of the nineteenth century and especially toward its end, the United States had tried to establish or strengthen its cultural, commercial, and diplomatic relations with European empires from the Ottoman in the East to the British in the West.[66] All the non-American cities asked about in this puzzle are either colonial possessions of one of those empires or important centers of commercial or political relationships between the metropole and its peripheries.

Pied puzzles engaged with world geography on two levels. First, by merging the imperatives of literacy with fun, they depended on the order of the

> No. 4.
>
> HIDDEN RIVERS.
>
> 1. Did they recover from their shock? 2. John, I left a package at the store. 3. George, do not tarry on the way from school. 4. You can't tag us, James, even if you try. 5. Oscar, Edith, and I are cousins. 6. Hem baby's dress before you leave, please. 7. He broke the large vase. 8. He got a rap on the head.   B. C. S. R.

FIGURE 3.11 "Hidden Rivers," *Harper's Young People*, March 11, 1890 (courtesy of Library of Congress).

letters of the alphabet in spelling real names that have an external, geographical reference point on the surface of the map or in the world atlas. Second, as a geography-based puzzle that had picked the names of quite a few significant American and non-American urban centers across the globe, centers that had either established or growing relationships with each other, they connected the young U.S. empire to the rest of the global imperial network. By connecting the American and non-American cities in one list, Ralston Seymour's puzzle was able to hint at the comparable importance that these cities had in world commerce at a time when the United States had begun to actively look for markets outside its borders and was especially seeking a footing in the Chinese market, which later led to its 1899–1900 entry into the Chinese market, amidst old, interimperial rivalries, through the "open door" notes issued by the United States to those European empires that held possession or short- and long-term leases in various parts of the Chinese Empire.[67]

In a similar manner, a puzzle printed in the March 11, 1890, issue of *Harper's Young People* (figure 3.11), hid the names of rivers in random sentences. "John, **I le**ft a package at the store" (emphasis mine), for instance, hid the name of the most celebrated river in what was then considered the "Orient." Encompassing names in random, otherwise straightforward statements that could have been uttered by children and adults alike again had to do with recognizing letters of the alphabet as capable of engaging in the act of spelling meaningful units as individual words on their own. To solve these puzzles, children had to first examine each letter of the alphabet individually and in isolation from other letters in the same or adjacent words, and then try to discard the nongeographical content of the statements in which they appeared in order to work out new words based on the order of individual letters, rather than according to the meaning of the whole statement.

Demonstrating stylistic flourish and spatial creativity, these puzzles underline the externally patterned, provisional, and dynamic cognitive maps that American children had in their minds while composing puzzles. Solving such geography-themed enigmas entailed two levels: first, guessing the geographical names that have the same letters of the alphabet in them, and second, correctly spelling the names of places near and far, perhaps with the help of atlases or geography primers. Finding the correct answers to this and similar puzzles could have reminded American children of the significance of international relations between the United States and the world by highlighting only a select number of dots on the world map with which they were more or less familiar. These enigmas suggest that literate American children played with letters of the alphabet in order to come up with not-so-arbitrary lists of geographical names that had geopolitical as well as commercial significance behind their seemingly neutral surfaces. Although hidden in banal, everyday statements about topics ranging from advisable child behavior to family relations, geography had become a constant presence in puzzles created by children. Moreover, even when shared with the magazine's readers as a means to challenge rather than to enlighten and to confuse rather than to direct, and even though they certainly replicated external, factual traits of space through an assortment of memories and representational tensions, these maps remained mostly unconstrained by them.

STUDYING THE PUZZLES THAT CHILDREN sent to *Harper's Young People* at the height of puzzle and game hysteria in the United States at the end of the century, the present chapter has discussed the ways geographical puzzles labored both to entertain children as they lost themselves in world maps and to let children show off their geographic knowledge of the world.[68] While not reflective of each and every child's cognitive map of the world, and despite the fact that no uniformly discernible thread can be detected between the place names that various puzzle makers chose to focus on, these puzzles on the whole point to rough, collectively drawn drafts of a dynamic dot map of the world in its relation to the United States and as imagined by a group of literate young Americans. Important to note is that it is hard, if not entirely impossible, to tell children's play apart from their more conscious imaginings of the world and to detach the labor that went into their designing and solving puzzles from the fun they might have had while engaging in such practices.

This is true of geographical facts and fantasies learned at school or through exposure to other media, namely travel, family connections, news, a specific country's shape or size, works of fiction, or simply the sound of the word that names a place.[69] However, children's puzzles that name American children, their food, and their pets after places of significance within the United States, or those that invoke place names beginning with letters of the alphabet that resonate with American values and identities, hint at a number of valuable conclusions to be drawn from this children's repository of American geographic knowledge. I outline five of them here.

First, the alphabet- or word-based geographical puzzles that *Harper's Young People* printed appear to have provided both their child creators and the children who tried to solve them with the opportunity to dissect drafts of internalized cognitive maps of the world and to translate them not into smaller pieces of the world map (as was the case with dissected maps) but into geographical place names. In doing so, these young puzzle makers would act selectively—if perhaps randomly—as they stitched together different places, consulted differently scaled maps, generated their own multiple cartographic scales, nested the toponyms they asked about in varying geographical divisions, underscored culturally striking connotations, and underlined geopolitical patterns. Despite their close resemblance to school primers' emphasis on world geography as a multitude of home geographies, the puzzles' overall location-distribution patterns did not follow what could be found in a majority of common geography primers of the time. Late nineteenth-century primers usually started with the United States and the Americas, and then moved on to teach about Europe, Asia, Africa, and finally Oceania, in the same or subsequent volumes. Children's puzzles, by contrast, bounced around the globe. Thus, chances are slim that any one of these puzzles was composed based entirely on geographic knowledge acquired directly through any systematic school geography curriculum. In sum, the geographical puzzles printed as Puzzles by Young Contributors involved far more eclectic dynamics and pursued far more diverse objectives than school geography books, geographical games, and dissected maps. Even seemingly simple puzzles, such as those previously discussed, did not follow the rule-based, hierarchizing principles that were, for example, found in the process of making dissected maps.

Second, unlike the common eighteenth- and nineteenth-century trend, noted by Michel Foucault, of bringing order to the Western drafts of geography and of hierarchizing the world in recognizable, exclusively Eurocentric categories, late nineteenth-century American children treated geography first and foremost as a site of literacy.[70] Answers to the numerous geographical puzzles they composed did not reflect the orderly lists of comparable parts of the world that would usually go together in identical scales that made a meaningful graphic whole, such as a map of the world or a table globe. And they did not, in the majority of cases, reflect the national and imperial order pursued and advocated by American adults. In the same vein, the young readers of the magazine had by this time been literate enough to come up with unprecedented imperial and transnational associations among letters of the alphabet, objects in the world, and subjects such as geographical place names—practices with which they could turn the world into an arena to wonder about, enjoy, and win over.

Third, the contents of these puzzles suggest that questions of world geography were extraneous yet expedient to the American child's play as long as they served the dual purpose of challenging other children and showing off both story-telling skills and knowledge of world geography. The naming acts involved in creating a puzzle suggest how mountain peaks, towns, and isles both near and far were made relevant to the young puzzle makers—as well as to those who would try to solve the puzzles—by bringing those places into the familiar boundaries of home and turning them into the constituents of an American family. In other words, these story puzzles suggest how the late nineteenth-century American family, its hope and future (its children), its means of subsistence and source of consumption (its food), what it looked after (its pets), and, more generally, the nation's complex definitions of and troubled distinctions between domestic and foreign, indoor and outdoor, and home and abroad, was at the same time an assortment of geographically marked curiosities.

Fourth, a word on children's sense of belonging to an unseen collectivity called *nation* and on where that nation stood in a colonial network, or family of nations: ever since nation-states began to become the norm in the eighteenth century, children have been born into nations (etymologically, *nation* involves the concept of birth) as well as into their families and into their given names. Christopher Kelen and Björn Sundmark rightfully argue that nations remind their citizens of their demanding presence through multiplying their signs and hiding them in plain sight to be invoked as quickly as is desired in times of national crisis.[71] "To the child, however," they assert, "the signs of nation [like the flag or the nation's map] cannot be buried in the unconscious; to get there they must first be to some extent known."[72] To take their conclusion a step further, as the puzzle-making practices in this chapter reveal, it is imperative to be cautious about viewpoints that consider nationhood simply as a collective project orchestrated by adults. Children themselves invoke national signs, like the national currency, the anthem, and the nation's history, as well as its geography. They consume these signs casually but earnestly, express curiosity as to their whys and hows, and at times resignify them in playful ways. For instance, when white American children grew old and literate enough to spot their given names in the evolving colonial contexts that ran over the map of the world, they did not consider the world to be a mere extension of their home, that is, their nation. Rather, they conceived of the world as a board on which others' homes neighbored one another and which stitched together seemingly irrelevant nations, connecting their home to the world, while also interweaving the world into complex, intimate geographies. Therefore, geographical puzzles were an invocation of the global, or at least the international, in the form of the ideal that geography was believed to be in its modern American sense: a collectivity of homes.[73]

And finally, examined from a contemporary perspective, these puzzles confirm that the "global" was an eclectic, partial, and mostly inconceivable scale for children. While the young puzzle makers attempted to put names from around the world in an orderly manner in alphabetical lists, their pretense at global exhaustiveness would fail to the extent that global—though nowadays a culturally and politically normative adjective—was, and still remains, an essentially idiosyncratic and contested term. Better said, in contrast to fact-based school geography primers, which operated on a solid basis as they went over continents and bodies of water in their aspiration to cover the entire world, children had their own personal terra incognita and their own time-bound lists of bright points on an imaginary spatial grid—their own cognitive maps of the world, that is—which they called on when thinking globally. As I argue in chapter 4, children's letters are yet another space in which they reflected on the commerce between the national and the global, observed and made sense of the world, and navigated potential paths to an adulthood of geographical privilege.

CHAPTER FOUR

We Sing a Geography Song
The Writing Child, the Portable Home Front, and World Geography

> Hubard [sic.] and I go to kindergarten. We learn a great many things. . . .
> We sing a geography song, about the continents, capes, isthmuses, peninsulas, rivers, and islands.
> —Maude R., *St. Nicholas*, October 1890

As the brief letter by the six-year-old Maude suggests, turn-of-the-century American children further took part in the national quest to mark the United States—their home—on the world map through the practice of letter writing. In this final chapter, I am drawn to read through the letters that children sent from the United States and abroad to the two most popular American juvenile periodicals of the time, *St. Nicholas* and *Harper's Young People*, during the 1890s. Similar to chapter 3, my arguments here revolve mostly around an exploration of the ways these young letter writers related to both American and non-American worlds/spaces first and foremost through practices in home geography.[1] In a manner comparable to the ways that children's puzzles grappled with home geography in a predominantly idiosyncratic manner, in children's letters, too, the practice of world geography found itself in conflict with children's tendency to engage with the views of adults, such as parents and relatives, by whom they were more immediately surrounded. At once personal and imitative, their letters document their encounters with and their perceptions of the world in its incomprehensible expanse and in contrast to their immediate, everyday life experiences around where they lived. As I argue in the following pages, this meant that the letter-writing child grappled with the "home" and the "world" elements of late-century home geography in different, at times contrasting, ways. While in some letters children labored to picture those who were allowed within the borders of home, in others they extracted the idea of "home" from its physical place in North America and carried portable copies of it with them wherever they went.

Long before they were typeset, children's letters were edited by adult mediators in a myriad of direct and indirect ways. This formal and thematic redaction process would begin with the selection of messages from a large pool

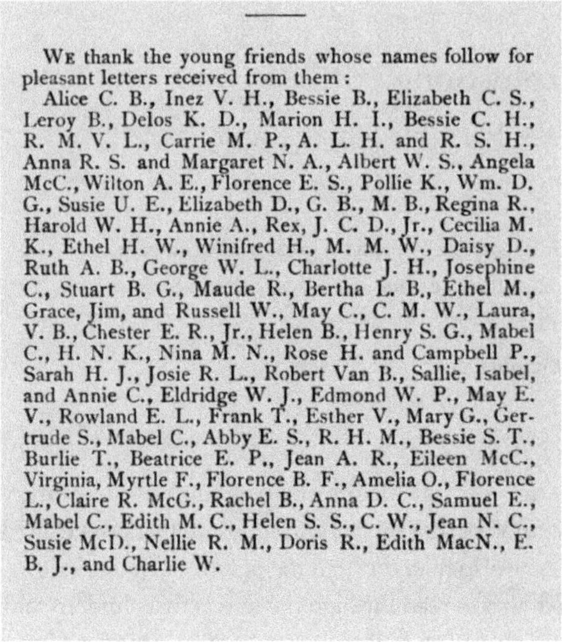

> WE thank the young friends whose names follow for pleasant letters received from them:
> Alice C. B., Inez V. H., Bessie B., Elizabeth C. S., Leroy B., Delos K. D., Marion H. I., Bessie C. H., R. M. V. L., Carrie M. P., A. L. H. and R. S. H., Anna R. S. and Margaret N. A., Albert W. S., Angela McC., Wilton A. E., Florence E. S., Pollie K., Wm. D. G., Susie U. E., Elizabeth D., G. B., M. B., Regina R., Harold W. H., Annie A., Rex, J. C. D., Jr., Cecilia M. K., Ethel H. W., Winifred H., M. M. W., Daisy D., Ruth A. B., George W. L., Charlotte J. H., Josephine C., Stuart B. G., Maude R., Bertha L. B., Ethel M., Grace, Jim, and Russell W., May C., C. M. W., Laura, V. B., Chester E. R., Jr., Helen B., Henry S. G., Mabel C., H. N. K., Nina M. N., Rose H. and Campbell P., Sarah H. J., Josie R. L., Robert Van B., Sallie, Isabel, and Annie C., Eldridge W. J., Edmond W. P., May E. V., Rowland E. L., Frank T., Esther V., Mary G., Gertrude S., Mabel C., Abby E. S., R. H. M., Bessie S. T., Burlie T., Beatrice E. P., Jean A. R., Eileen McC., Virginia, Myrtle F., Florence B. F., Amelia O., Florence L., Claire R. McG., Rachel B., Anna D. C., Samuel E., Mabel C., Edith M. C., Helen S. S., C. W., Jean N. C., Susie McD., Nellie R. M., Doris R., Edith MacN., E. B. J., and Charlie W.

FIGURE 4.1
List of children's names whose letters were not printed, *St. Nicholas*, May 1894 (courtesy of HathiTrust).

of incoming letters. During the 1890s, *St. Nicholas* and *Harper's Young People* each published an average of ten letters in every new issue. It is evident that they received more letters than they had space to print—a fact that the editors acknowledged by printing a list of names of those children whose letters were received but not printed at the end of almost every The Letter-Box (*St. Nicholas*) and Our Post Office Box (*Harper's*). These lists suggest that about one out of every ten letters sent to the two journals was printed (figure 4.1). Despite—but also as a result of—the editors' touch, the letters reflected continuity, as well as aberration and novelty, in both form and content. Consequently, these letters present a vibrant space in which intergenerational exchanges in encounters with the world can be traced and examined. At the same time, despite their brevity, they form a historically significant collection of "writing tasks" in which children practiced translating their abstract mental maps of the world into concrete words that they shared with their unseen, unknown readers. In these writing tasks, children dispensed with list making and storytelling and made sense of "home" and its inverse image of "non-home," ultimately to arrive at definitions of home and what and who, in their view, belonged to it.[2] At the same time that these, occasionally cripplingly edited, letters do not reveal all that American children thought or felt about the world, they do point at patterns behind the ways child writers merged what they had learned in school geography with what views and emotions their encounters with spaces near and far prompted in them.

Not surprisingly, a great number of the letters printed in both *St. Nicholas* and *Harper's Young People* were written by children living in different parts of the country. These "home-made" letters contained information about their writers' physical appearances, siblings, pets, toys, and favorite books, but also about where they came from in the United States. They entailed accounts of their summer trips to the coast or their fishing adventures in a pond nearby. An overwhelming majority of them provided lists of their Christmas presents, enumerated the subjects they studied at school—mainly geography, history, English, and arithmetic—and mentioned their favorite writers and stories in the magazines. American children who had traveled or lived abroad, with or without their families, made their voices heard by writing letters from near and far. Their letters provide us with a glimpse into how they viewed different parts of the world as they experienced them firsthand. We can see, for instance, in which parts of the world they felt the urge to express awe, curiosity, or disdain about the local culture or where they amused themselves with the idea of describing "exotic" scenery, customs, and peoples to ultimately undergird their Americanness while away from their U.S. home. In sum, the repository of children's letters furnished by the runs of *St. Nicholas* and *Harper's Young People* provides an essential supplement to the publicly available documents that were generated solely by adults, such as books and maps, and furnishes evidence, however mediated, as to how children perceived nationhood and national spaces as a home among many.

Children's Magazines as Accidental Archives:
Editors and the Writing Child

Justifying the necessity of launching a juvenile monthly with Scribner and Company, Mary Mapes Dodge spelled out her views regarding childhood and the reading child in "Children's Magazines," anonymously published in *Scribner's Monthly* in July 1873. Warning that European, especially British and German, juvenile periodicals "distract [children's] sensitive little souls with grotesquerie," Dodge then turned to American children's periodicals: "We edit for the approval of fathers and mothers, and endeavor to make the child's monthly a milk-and-water variety of the adult's periodical. But, in fact, the child's magazine needs to be stronger, truer, bolder, more uncompromising than the other. Its cheer must be the cheer of the birdsong, not of condescending editorial babble.... A child's magazine is its pleasure-ground."³

Children's letters belonged to a pen-and-paper corner of this "pleasure ground." Informed by these views, epistolary spaces of interaction between young letter writers, adult editors, and their young readers constituted a constant feature of both *St. Nicholas* and *Harper's Young People* from their very first issues.⁴ Unlike the more private letters that both adults and children commonly exchanged with each other, "fan letters" by children were in fact part of a larger public epistolary tradition with the power to satisfy, at least partially, the emerging American middle class's desire to enter public conversations that had for long been going on exclusively among the elite.⁵

Since *St. Nicholas* was first published in 1873, letters by readers had been one of the magazine's columns. This had allowed Mary Mapes Dodge, *St. Nicholas*'s editor during its first three decades of publication, to develop strict yet receptive editorial policies regarding which letters were printed. *Harper's Young People*, too, launched its Our Post Office Box column in its first issue with a call for letters.⁶ Following the initial call, the column started with two very short letters in the second week of its publication—that is, Tuesday, November 11, 1879: Willie J. H. wrote a letter in which he asked a question about his pet alligator, while Lulu W. inquired after the possibility of establishing a "work-box department for little girls" at the magazine. Over the years, with the ultimate power to decide which letters were printed and to what degree they should be edited beforehand, adults would set the standard for "printable" letters. Put differently, the prototype for the children's public letters under study here was built around the examples provided and the expectations set by the letters that had already found their way to The Letter-Box and Our Post Office Box by editorial decision.

As more letters were received, the "Postmistress" started the first Our Post Office Box in 1880 to suggest topics about which its young readers could compose letters: their pets and their observations of nature, including birds and flowers.⁷ A decade later, American children were writing letters about numerous other subjects, ranging from the color of their eyes and hair to their views on the Spanish-American War, and from what they thought of non-Americans to their encounters with migrants in New York. Regardless of what they wrote about, writers of a great majority of these letters offered cautious pretexts for writing a letter, which they hoped would deserve to be printed. Some wrote second or third letters in the hope that this time their letter was printable. Sometimes children also wrote letters that directly addressed the larger epistolary community of which they found themselves a part. In her letter to *Harper's Young People*, dated July 22, 1890, Kate B. Greene asked two other girls whose letters had been previously printed in the magazine to write again and say more about their country—that is, New Zealand. In a similar instance, Joseph C. Henlings (*Harper's Young People*, July 22, 1890) and C. Taylor (*Harper's Young People*, January 13, 1891) wrote in their letters that they wished to exchange stamps with foreign children. Ultimately, writing and getting printed took place in an environment predominantly regulated by adults.[8] Yet how and what children wrote varied depending on their age, gender, writing skills, and immediate life experiences, meaning that their individual voices and experiences did, to some degree, show through the editorial veneer.[9]

The Writing Child at Home: Earmarking Self and Other

Many children sent letters from across the United States in attempts to sketch "home," share with their readers its geographical and emotional coordinates, and catalog and introduce its residents. These attempts involved combining the political and the human elements of modern geography with the range of emotions children felt and of impressions they had formed toward what legitimized their belonging to this or that home. Conceived at this intersection as writing exercises in home geography, we can approach their letters in search of the answers to an imbricated number of questions: What/Where qualifies as home? Who lives in it? Who has the potential to join its expanse in the longer run, and in what capacity? What sets home apart from the void, the "non-home"? And what degree of cultural traffic does it, as a space characterized by familial order, geographic liminality, and civilized self-containment, allow with the allegedly chaotically uncivilized, confusingly ungeographical, ominously encroaching void?

Looking for answers to these questions, we are confronted in these letters with children who, through attempts at individual self-portraiture and with the help of multiple acts of inclusion and exclusion, insisted on belonging to the nation as their imagined but rightful home: the liminal national space, the home-of-their-homes. Moreover, read together, these letters suggest that a thorough and precise mapping of the nation-as-home hinges on how the nation itself can be divided up into a litany of smaller homes and the handful of "non-homes" (such as reservations, prisons, and quarantine stations) that filled the distances between smaller home units and walled the nation off from unchecked waves of immigration from outside its borders. In order to better make sense of an imagined community as large and eclectic as the

United States—the home-of-their-homes—as inclusive of both homes and non-homes, it is perhaps helpful to reexamine the nation's geographical division into these units in terms of racial compartmentalization—what Michel Foucault calls biopolitically induced "caesuras":[10] homes as spaces of inclusion that both map and shelter within their boundaries those individuals who have the right to citizenship, political enfranchisement, and sexual intercourse and reproduction, versus non-homes as spaces of exclusion or in-betweenness to which the inferior, undesirable Other is expelled (as in the case of reservations), or otherwise detained (as in the case of quarantine stations or detention centers at borders).

Almost anxious to establish their right to belong, many children began their letters by establishing that they belonged to households where they were one of several siblings. This was followed in a lot of letters with brief or elaborate acts of self-portraiture: the color of their hair and eyes, their freckles, and so on. At the same time, none, during the 1890s at least, referred to their skin color. Altogether, whiteness as a racial category of privilege and as one of the "foundational myths" of the nation throughout the course of the century seems to have presented itself to this group of children as a "natural" sign of belonging to a nation that was collectively, categorically, and more or less unanimously blind to other colors and Others' colors.[11] Far from abolishing whiteness, the absolute absence of references to skin color as a category in children's attempts at self-portraiture underscores the essentialized and internalized understanding of whiteness as unmarked but universal. Indeed, since whiteness was the enforced norm, it was other colors that were pointed at as unnatural birthmark inscriptions on the bodies of Others. Consequently, it was nonwhite letter writers who found themselves under pressure to watch out for how they described their looks—indeed their bodies. After all, for them, mentioning the tone of their skin could have totally undermined the legitimacy of their right to residency in a "home" that was deemed white.[12]

In the same vein, none of the printed letters ever mentioned the child writers' European, African, Native American, or Asian ancestors. None boasted about their family roots or their remaining relatives somewhere off the map of the United States. At the same time, although some children mentioned taking German or French as foreign languages at school, none referred to any language other than English as the spoken language at home.[13] Without a doubt, by the end of the century, the young subscribers to *St. Nicholas* and *Harper's Young People*—and their families—had already embraced English as the language of the nation and the family. Indeed, children's parents and teachers had already signposted the borders of their "home" in English, further sealing them with literacy, whiteness, physical health, and hygiene.[14] Furthermore, at the turn of the century more than ever before, white Americans—including the magazines' editors—agreed that being truly American meant having left Europe behind. This meant that even for those children who *did* speak another language at home, social prestige and the negative reinforcement they received at school and through previously printed letters in their favorite magazines would surely have discouraged them from confiding such facts about their family lives in public.[15] In other words, all the parties involved in giving shape to this epistolary space made sure that there was no question about letter writers' "Americanness."[16] In the end, while the nation indexed *their* identities, children, too, had learned to reflect on the nation as the home of a predominantly white Anglo-Saxon populace.

Busy mapping themselves on the expanse of this nebulous national home, many young letter writers argued that their letters deserved to be printed not because they were well-mannered, literate, white American children who had mastered English but because they wrote from small towns or other more sparsely populated regions from which they "have not seen many letters" (Grace W. K., *St. Nicholas*, September 1890). In effect, pinning the geographical regions they came from onto the mental map of the magazines' other young readers, children tried and mostly succeeded in ensuring that they wrote as representatives of their hometowns: "As I have never seen a letter from here," wrote May F. B. (*Harper's Young People*, January 20, 1891), "and as I have from nearly everywhere else on the globe, I thought I would let you know there was such a place as Lockport, Illinois." In a sense, these children identified themselves as mappable members of the community they came from or the community where they lived at the time they wrote a letter and therefore communicated their locally pinned sense of national belonging. Evidently, for a majority of them, it was not membership in the nation but their belonging to a specific town or region that signposted their hometown on the map of the United States and, by extension—in the case of children like May F. B., at least—on the map of the world.

But who else, other than them, belonged to, or was welcome to stay in, this home? During the period under study, I did not encounter a single letter in either of the periodicals that mentioned foreigners, new immigrants, or economically underprivileged nonwhite Americans as having been present in spaces near where these children lived, played, and received schooling. Overall, whether these children were from Boston or Zanesville, it is as if an invisible curtain was cast by adults between children of well-off families and children of former slaves or newly arrived immigrants. In fact, only occasionally did a letter get printed in which young American children mentioned encounters with "others." Laboring to identify and further mark the boundaries of home, an occasional longer letter would hint at their authors' encounters with foreigners and with newly arrived immigrants, and then only in exceptional or adventurous settings and in spaces that were clearly marked as separate from the communities where the letter's author lived.

Some letters gave lengthy accounts of Native Americans, reproducing them as "semi-savage" curiosities who occupied the undifferentiated, outlying "void" between the homes the nation encompassed. Having inhabited the Americas long before the United States began to form as a nation (and, in the case of African Americans, having been forcibly brought to the country), members of these communities were victims of a strong historical drive in the young republic—"the power of imposed identification"—through which privileged white people tried to erode the identities of entire communities based on factors that would reinforce the power hierarchies on which white privilege hinged.[17] Therefore, in composing letters about their unfamiliar food and customs and their allegedly warlike behavior, children referred to the actual inhabitants of "Indian Reservations" only occasionally. While writing about life in "Indian Territory," Maggie Taylor's letter, printed in *St. Nicholas* in May 1898, declared that white people—by which she meant white children—"never see the real Indians." Maggie voiced concerns about her safety, however, as if spatial proximity to Native Americans was synonymous to danger or an inherently vile experience, whether or not she had any face-to-face encounters or exchanges with them. In another instance, Guy P. B.'s letter (printed in *St. Nicholas* in January 1894) gave an account of the Oneida among whom he lived with his missionary father and family. Elevating himself to the powerful position of an adult in relation to the infantilized Oneida and parroting an egregiously racist, white supremacist adult rhetoric, Guy bragged: "We have grown fond of them during this time for they are such simple, childlike, lovable people." In Guy's words, thanks to the evangelizing activities led by people like Rev. Goodenough and Guy's own father, the Oneida were no longer a "war-like" tribe that was "feared among all their neighbors" (before they were removed from western New York and forced to resettle in Wisconsin in the 1830s). Today, he rejoiced to mention, they were a community of "grown-up children, [who] like to be treated like men and women."[18]

In a comparably longer letter that was printed in *St. Nicholas* in October 1891, Hope G. gave an account of her life as the daughter of a missionary family in "the agency of the Mescalero Apache Indians." Hope's letter entails a rather detailed account of five years of forced schooling of Apache children under her mother's educational grip as the founder of the reservation school. Following two centuries of harsh practices systematically implemented by white settlers to displace, forcibly Christianize, and later "Americanize" Native American children, Hope's mother's boarding school stood witness to a new "Indian policy." In effect, by the 1870s, and particularly in light of the enforcement of the 1887 Dawes Act, which promised to allot land to individual Native Americans in order to convince them to settle for farming and to give up their tribal identity in exchange for U.S. citizenship, the U.S. government had begun to realize that educating Native Americans from a young age and away from their tribal land was "the cheaper and more constructive approach to assimilating young Indians."[19] Hope's parents were agents of this nationwide policy that, in the name of education and despite objections and resistance by Native American parents and children, intruded in Apache homes and intervened in their culture.[20] As Hope complained in her letter, "It was difficult to get parents of the little Indian children to consent to their coming and living in the building provided for them." The solution, she reported, was violence: "At last, the agent advised the plans of sending out and forcibly bringing the boys in, where they were subjected to a bath, much against their will." The white missionary family's efforts to snatch Apache children from their homes and to estrange them from families and culture would find new, affective dimensions as they changed Native American children's appearance: "With their long hair shaven off, and a suit of American clothes substituted for the Indian apparel," Hope rejoiced, "the transformation was in some instances so striking that it was hard for their parents to recognize them."

Nonetheless, this sweeping, simultaneously physical and psychological aggression did not end with mastery over the boys' bodies and brains. As Hope observed, her mother "would provide them with an American name in full, such as Philip Sheridan, Miles Standish, Christopher Columbus, or any that suggested itself, and they were always known by these names afterward, dropping their Indian names."[21] Changes in their appearance, education and socialization, and names—names as bitterly ironic as Christopher Columbus—were part of a larger two-sided project that had defined Native–white settler interactions for well over two centuries: On the one hand, through educating, transforming their looks, and giving them new names, white missionaries and teachers intended to fully rewrite the Native American boys' identity, erase their communal culture against their will, and force them to emulate white culture with the fake promise that they would begin to belong to the nation. On the other hand, white Americans did not offer Native Americans equal rights or privileges based on the claim that, since they were inherently and visibly different, their right to join the national household had to be deferred. Though pushed off their land and deprived of their home and heritage, they were marked as—and were to remain—hyphenated Americans, unalterably distant and irrevocably foreign. While their homes were violated in the name of progress and national unity, Native Americans were not given a new home.[22] Displaying a great degree of conceit and indifference in the face of such a flawed and cruel project of displacement (forcefully taking Native American children away from their already displaced parents on reservations), Hope celebrated this double displacement as the success of the white American missionary in regulating who belonged to the interiors of the allegedly monoracial American home and under what terms.

As proof of the beneficial role her mother had in fostering this nationalist project, Hope attached to the end of her colorful letter yet another letter that James La Paz, one of the "little savages" from the La Paz tribe, had written to her after her mother had sent him to an "Indian school" in Grand Junction, Colorado. The *St. Nicholas* editors decided to print James's letter "as it was written," that is, without the customary editing that other letters received in terms of content and language:

TELLER INSTITUTE, GRAND JUNCTION, COL.

My dear friend, MISS HOPE G.: I have received you letter some time ago, and I was very glad that you still remember me. I don't think you are a lazy girl, because you wrote a very long letter. When I first wrote to you I think you were off somewhere; that is why you did n't write me soon. No, I never get mad at anybody when I don't get letter from him soon. I study arithmetic, fourth reader, and some time geography, but not very often. When I saw the writing on the envelope I did n't think it was your writing. I thought it was from my home, Because it look like baby writing. Can you write better than that? I think you can if you just try. Our teacher is gone home about a week ago. I was very sorry when my teacher is gone, that I pretty nearly cried for her. She treat me better than the other boys. She gave me one of her picture and book. She use to give me candy every day.

Hoping to hear from you soon,

Yours, James La Paz.

Here are the facts: A Native American child writes a private letter to his white friend (who also happens to be the daughter of his schoolteacher). The white friend attaches the letter to her own public letter and sends it to be printed in one of the nation's most popular children's magazines "as it was written." It appears as an appendix, a postscript, in a juvenile monthly that James does not have access to. His letter ends up being framed by the editors not as an achievement by a literate young boy (or even as proof of a much-desired "progress" among his people) but as a control sample of poor writing. In deciding against revising James's spelling mistakes and faulty grammar and punctuation, the editors seem to have taken the opportunity to advise their readers to practice and master the art of epistolary communication in order to avoid embarrassing themselves in a public platform such as a juvenile periodical the way James La Paz—barely literate and not belonging—had. Printed as a cautionary tale, James's letter was adopted to warn Hope's unseen white friends across the nation that to write poorly was "baby writing," a fact which, even James seems to have believed, was to be expected *only* of very young children and, by extension, of "childlike" Native Americans. In the end, in privileging the natural flow of language over form, it was James who cautioned Hope and other American children about the ways a letter should *not* be written.

Associating foreignness with poor writing skills, James was maximally othered with both his faulty English and his adopted (in fact, imposed) name, which appeared in full at the end of his letter. It was as if—unlike white children whose names were usually followed by their initials and not their full family names—it did not matter in the case of an outsider such as an Apache child if they were to be fully identified. After all, displaced and renamed as they were, knowing Native American children's full names would not mean that private details about their families (such as where they lived or who their parents were) could be sleuthed. What is more, the real erasure done to James's identity was taking place not at the level of his "given" name but in fact at a site far more intimate and deep-rooted that that: his place in the fabric of a nation that gave Hope the right to a home and James the right to a reservation.[23]

Matching Hope in tone, eleven-year-old O. T. H. wrote a letter to *St. Nicholas* to describe the immigrants' quarantine on Staten Island, off New York Harbor. In unusually accomplished prose for a child of that age, O. T. H.'s letter (printed in the magazine's October 1890 issue) mentioned a visit "to Staten Island and to a place called Quarantine, just this side of Fort Wadsworth, where there is a beautiful view up and down the bay." The quarantine station that this child had visited functioned under the direction of the New

York City Department of Health. In fact, since 1799, the Marine Hospital Service on the island had been responsible for isolating newly arrived immigrants who were infected with cholera and other contagious diseases from the American population. Staten Island—together with Hoffman Island and Swinburne Island, which, respectively, treated smallpox and yellow fever—was the site of the very first contact a majority of newly arrived immigrants had with the United States. It was, on one level, a spatial incarnation of the non-American world, migratory fragments of the world that had reached the shores of the United States, and, on another, a spatial materialization of the notion "foreign to the United States in a domestic sense."

This oxymoron was first appropriated in the fiery 1901 legal discussions in the United States Supreme Court in *Downes v. Bidwell* over the potential annexation of Puerto Rico and the citizenship status of Puerto Ricans within the domestic expanse of the U.S. empire.[24] The phrase was used to argue that though residing on land that now belonged to the United States, and thus legally bound to the nation (that is, domesticated and nationalized), Puerto Ricans could not be considered citizens, because they as a people had been deemed foreign to it. As such, the "new" people on U.S. land—including Puerto Ricans, Filipinos, and immigrants who had just landed in the quarantine islands—were, at best, assigned an in-between political identity that encompassed two extreme ends: physical presence *and* legal absence. To return to O. T. H.'s letter, then, both Puerto Rico and the quarantine station, as metaphorical "unincorporated territory[ies]," were foreign to the United States in a domestic sense, both "confound[ing] the borders between the foreign and the domestic."[25]

In O. T. H.'s words, based on what they had probably heard while touring Staten Island, "quarantine means forty days, because ships that had contagious diseases like yellow-fever or small-pox were kept away from the land forty days; but now it is different." In their account of the quarantine, which was strictly observed during the last two decades of the century, the separation started as soon as an immigrants' ship reached New York Harbor: "When a ship comes into the port it is ... telegraphed to the Quarantine station. When the ship is seen coming around the fort, a bell is rung for a health officer; he goes on board a steam tug or sometimes in a rowboat to the ship and all the emigrants pass before him, and if there is on board the ship any contagious disease to which passengers have been exposed they are taken off and the ship is fumigated."

According to O. T. H., the cause of such contagious diseases is "a little animal which I think is called microbe, which makes people have yellow-fever and it sometimes gets into clothing, so all the clothing is put in an oven in a furnace where all these little animals are killed by heat or steam." While fascinated by the fate of the microbes, the letter remained entirely indifferent to the immigrants, their fate in the quarantine station, and—if they survived the quarantine period—their life in their new home in the United States.

The quarantine station had a far more complicated backstory than O. T. H.'s letter reported. Since late seventeenth century, quarantine had been observed as one of the first measures taken in regulating U.S. borders every time an immigrant ship was about to enter New York Harbor. Meant to control the outbreak of contagious diseases like yellow fever (which, in the case of the Philadelphia epidemics of 1699, was believed to have been brought to the North American continent from the Caribbean), the quarantine practices were at first locally organized. The 1890s witnessed a number of anti-immigrant practices that made the passing of the National Quarantine Act of 1893 a controversial prospect on the national, rather than the local, level.[26] According to this act, newly arrived immigrants and the crew members aboard those ships that had been struck by a contagious disease were supposed to be taken to one of three islands off New York Harbor and be subjected to a quarantine period in one of the more or less specialized hospital buildings. Healthy immigrants would spend the quarantine period in residences on the island. While first-class passengers were supposed to reside in the hotel-like St. Nicholas building on Staten Island, second- and third-class passengers were put in shanties—buildings that O. T. H. might have seen from afar on their tour of Staten Island.[27]

Supported by U.S. laws and backed by popular sentiment, though debated among health activists, the quarantine station served as a space where the foreign—the undesirable, unhealthy Other—was literally kept "at bay," away from the supposedly germ-free, healthy American home. The very act of unclothing newly arrived immigrants on the spot—that is, dispossessing the sick Other—was heavily charged with racist and class-based prejudices and practices. At the same time, Americans who traveled abroad were also subject to usually shorter periods of quarantine, as was the case with Alice and Marion White B. from Houston, Texas—albeit under starkly different circumstances. Quarantine was by this time so unquestionably associated with separating the foreign from the native that when, on their way back from the Catskills during the "cholera scare" of September 1893, Alice and Marion learned that they, too, had to be quarantined, they were "very much surprised." Even then, their experience stood in stark contrast to the usual quarantine experiences of immigrants: "We had great fun riding backward and forward on the tug," wrote Alice.

They had been free to go bathing and fishing, the quarantine doctor had been friendly to them, and "he and the captain did everything in their power for us."

While foreigners' bodies were quarantined on the island, their clothes "were removed to be washed on the spot, carried to washhouses, or burned."[28] And like their clothes, a considerable number of these immigrants were not going to survive the quarantine period.[29] Deceased immigrants' bodies would be buried in a graveyard on the island before they had ever set foot on U.S. land. Until well into the twentieth century, the island continued to serve as a space of (dis-)entanglement, where unhealthy and unwelcome foreigners were to be cleansed before securing passage into the domestic sphere. Consequently, the idea of living on an island that was viewed as the site at which foreigners with contagious diseases were being treated, or simply kept quarantined but untreated, did not please the residents of Staten Island. They associated not only the quarantine buildings and the cemetery but also the Americans who worked at the station with foreignness and, subsequently, with disease. Although not linguistically collocational, foreignness and sickness were a common cultural pair in the eighteenth and nineteenth centuries. And just as sickness was thought to be a result of germs' intrusion into the body, foreignness was understood as the migrants' intrusion into an allegedly wholesome national body.[30] It was as if even the American personnel of the hospitals and residences were rendered foreign simply through tending to the quarantined immigrants.[31]

Although entirely beside the point for the young O. T. H., whose curiosity revolved around the fate of the microbes and the cleansing function of the quarantine station, the history of these immigrants and the biopolitical exercise of state power over their bodies in quarantine stations (and today in state-run and privatized for-profit detention centers) are part of the history of these islands as well as of all the other places in the country where immigrants would later settle. Remembering that immigration from Europe and Asia was the result of complex individual, familial, local, national, and transnational factors—including joining one's family or escaping ongoing political, economic, or environmental hardships as a result of interimperial rivalries, famines, or unsustainable depletion of natural resources—it should be noted that the history of seemingly localized quarantine stations is as much a part of the national history of the United States as the stories of the individual immigrants who arrived, waited, died, or left those spaces are part of the history of the nations and colonies from which they had emigrated. In other words, the history of such spaces inevitably interweaves—spatially as well as diachronically—national U.S. history with the biographies of all those individuals who reached its shores from almost all over the world. As Epple, Kaltmeier, and Lindner rightly assert, paying attention to such dense micro/macro entanglements is a useful approach to more thoroughly uncovering the trends, decision- and identity-making processes, (im)mobilities and tempos, and similarities and differences that connect individual biographies to the familial, regional, national, and global—and even, at times, the seemingly disconnected or unfitting—fabric of which history is variously made.[32]

Obviously none of these issues surfaced, even if tenuously, in O. T. H's recollections of the island's tour, most probably because as a whitewashed, carefully spun narrative about migration, meant to entertain the young children who toured the island, the tour informed them about germs and the procedure to defuse their biological threat to human bodies in a "child-friendly" manner that prioritized the purported accessibility of scientific processes. At the same time, it barred any quasi-adult, ideology-driven, supposedly austere, and nonconcrete insights into the life stories of the people who had been to the island or into the place's historically ambivalent function as an entry point to the United States. Ultimately, it is not too far-fetched to argue that quarantine stations did to immigrants something akin to what Indian schools did to Native American children: cut and curated foreignness, and put individual identities under erasure in the name of the exclusive and exceptionalist homogeneity that the American home exacted.

O. T. H. was not entirely alone in writing to their favorite magazine to report their encounters with or to record their emotions toward foreignness. In the April 1892 issue, the editors of *St. Nicholas* printed a letter by Gertrude Du B., who wrote from Hudson, New York. An outstanding example of children's interest in and deep-seated sense of anxiety toward foreignness, this letter included no information about Gertrude herself. Instead, it enclosed a "beautiful *patriotic poem*" called "War" that Gertrude's nine-year-old brother had composed. Reminiscent of O. T. H's letter in style, Gertrude mentioned in her letter that her brother wished that he had instead called the poem "'They are coming,' which seems more appropriate, there not being much war about it." Composed in the blatantly anti-immigration and nativist tone common in the 1880s and the 1890s, this poem allows a glimpse into young white Americans' view of their home, which they feared was surreptitiously under attack by immigrants:

> They are coming, they are coming,
>   To destroy our native land:
> They are coming, they are coming,
>   From every shore and strand.

They are coming in the morning, they are coming in the night,
And now, my fellow-countrymen, we must
all take flight.

They are coming, they are coming,
With all their swords erect,
They are coming, they are coming,
    Ourselves we must protect.
They are coming in the morning, they are coming
in the night,
And now, my fellow-countrymen, we must all
prepare to fight.

They're upon us, they're upon us,
    Oh, help us every one!
We'll be murdered! We'll be murdered!
    The father and the son.
And now we must prepare to flee
Across the meadow and the lea.

Deemed "beautiful" perhaps because it was "patriotic," the poem imagines three possible, equally dire, responses to encounters with immigrants: run away and let foreigners enter and occupy home; fight and protect the nation's borders; or do nothing and get killed. As we saw before, O. T. H.'s letter expressed amazement at the power of the quarantine stations in putting immigrants in an object position (*them*) by subjecting them to systematic debasing measures (undressing and examining their bodies, burning their clothes) in order, allegedly, to cleanse and cure them before they entered the national body. Chiming in with the nativist views of the adults around him, Gertrude's brother's poem, on the other hand, imagined his generation in an object position (*us*) and as threatened by immigration "From every shore and strand."[33] Coming "with all their swords erect," the poem warned, their sole purpose is to "destroy our native land." Repeating "they are coming" a total of twelve times in his short poem (in fact, more than a third of the poem consists of this phrase), he mimicked adults' alarmist, nativist rhetoric that must have informed his own understanding of immigration. After all, far from a coincidence, this letter appeared in The Letter-Box, the same column that, in the wake of heightened nativist sentiments in response to the Spanish-American War, the editors of *St. Nicholas* had decided was the ideal place to invoke patriotism in children by reprinting the full text of the "Star-Spangled Banner" (July 1898).

On the whole, American children writing from home imagined the United States as a homogeneous space bookended by quarantine stations (to regulate, stop, or at least slow down the inflow of foreign individuals into the territorial expanse of the nation) and sprinkled with a handful of reservations (and white-run boarding schools away from reservations). In their letters, children mapped quarantine stations (as well as reservations and tenement houses where newly arrived migrants lived) while they also singled them out as not belonging to the national stretch of home. In other words, they approached these non-American spaces as the image of the nation *ex negativo*. Curiously enough, no mention was made in their letters of any deterrent infrastructure at the Canadian and Mexican borders, as if they imagined that the United States occupied the entire length of the Americas on the map, stretching all the way to the north and the south, while it was bracketed to the east and the west by frontiers that separated Americans from non-Americans and maintained the sanctity of the American home by stratifying and spatially confining foreignness within tenement projects, reservations, or boarding schools.

The Writing Child Abroad:
Marking "Home" Away from Home

While at home, American children seem to have either experienced the non-American world as a homogeneous, undefined, and undifferentiated distant space that belonged in geography textbooks or imagined it vicariously as the inferior pair in an American–non-American dichotomy that had long been the subject of adventure books such as the Rollo series. Alternatively, they would dismiss the non-American world as undesirable and unfamiliar and banish those instances of it that reached the United States to quarantine stations, urban ghettos, internment camps, and so on. While abroad, on the other hand, young Americans would experience the world firsthand, moving from place to place as tourists who visited "non-home" historical or religious sites of attraction or as residents of makeshift homes "away from home." These letters provide us with an epistolary space in which we can look for the child in the world, with an imperative to examine their in-the-world-ness from various perspectives: How did children engage with the confusing territorial expanse of the world, smaller, metaphorical representations of which they had already encountered in books, maps, and map toys? In what ways did they exercise home geography or depart from it as they assumed roles other than students (such as supra-geographical roles as observer, tourist, expatriate, letter writer, and storyteller)? Engaging with this set of letters brings to the surface the ways that American children came face to face with the world: as literate, increasingly mobile young people who replicated American adults' evangelizing, commercial, or diplomatic activities in the world at large, or as sons of rich mothers who visited European spas to improve their poor health or daughters of missionary fathers who built a missionary outpost on the Chinese or Indian "frontier"—experiencing, appropriating, and mapping the world for themselves.

Brief as children's public letters had to be, such epistolary spaces were meant to furnish the socioeconomically less privileged children with the opportunity to explore the world on their free-of-charge journeys of imagination and to locate themselves in its expanse. All this could be done by simply reading the magazines, without the necessity of physically navigating the actual geographic distance between home and the near and far corners of the non-American world. Reading a letter printed in the May 1894 issue of *St. Nicholas* by Edith M. K., for example, children could form ideas about the far-off colonial India. According to Edith, in the middle of an elephant ride in the Indian countryside, Edith's missionary father had shown them "from that elevated position" what he had called "an Indian dinner party," in which "the guests were some vultures and jackals, who were feasting on the remains of a dead buffalo." On another occasion, Edith's father had biked "through a troop of monkeys, who were evidently very much surprised to see that new mode of locomotion." Mostly a summary of her father's views, Edith's letter visualized India for her readers by making subtle references to entities that resided on either side of the thin, porous line that her father drew between India's wildlife and its human population, calling wailing jackals "little gentlemen going to a concert." In the absence of letters in which American children gushed about an unforgettably great time in a home other than the United States or in which they complained about having felt inferior to the locals, the printing of letters such as Edith's are most illuminating, since they reflect the broader perceptions and imaginings of white America—supremacist narratives that were shaped not simply by children's family trips, free-time readings, and studies but also by what children saw firsthand.

Unlike Edith, most children completely omitted the human element from their letters about non-American spaces. Printed at a time when the United States was trying to exploit the islands' natural resources as its own, Sophie B. J., the daughter of the American chief justice of the Republic of Hawaii, wrote a letter (printed in the December 1895 issue of *St. Nicholas*) in which she listed various types of fruit that could be found on the Hawaiian

Islands. Apparently uninterested in the Hawaiian population itself, she instead turned her attention to the islands as an empty spatial appendix to home—a space where she could spend holidays and go horseback riding. Written in a strikingly similar tone, Edith H. B.'s letter to *St. Nicholas* (printed in the November 1890 issue) from Kohala, Hawaii, described a native food called Kalo. As if hosting a popular science show, she went to great lengths to explain how Kalo is planted, irrigated, and prepared for eating without once mentioning the people who had, for generations, planted, prepared, and eaten it. Mirroring American adults' century-long obsession with the Pacific and particularly the Hawaiian Islands as an integral part of "the story of American imperialism,"[34] children's exclusion of the Hawaiian people from the picture replicated the rhetorical displacement of Hawaiians within their own home even before Hawaii was officially annexed to the United States in 1898.[35] What is more, this erasure, racial and racializing, had the power to simply recast the islands that Americans had long coveted as an uninhabited wilderness, thus excluding Hawaii from the very frame of comparable homes in which American children had learned to think of the world. This was a simultaneously supremacist and reductionist exercise in writing off the human element from space—a practice through which children willfully digressed from home geography, with its emphasis on environmental and human elements. Such a digression raises a number of urgent questions: What is a home when emptied of its human residents? Who evacuates and reoccupies it before it recedes into ruin or is claimed back by nature? And, ultimately, in determining the known, or at least the potentially knowable, boundaries of home, who determines who or what should be abandoned into the uncharted "void" between homes?[36]

In writing about other homes and Others' homes, children's letters continued to return to literacy, educability, and assimilability as eligibility criteria for membership in the national household. An unsigned letter by an eight-year-old boy from Honolulu to *St. Nicholas* (September 1898) gave an account of a devoted Hawaiian servant. As the letter reported, Wai Waiole, the white child's companion, was a little boy. With a backhanded, hardly discernible hint at a possible homoerotic tension between the two boys, the young writer confided in the readers that "when I bathe I make him go away off behind the house, but sometimes I catch him looking to see if I'm all right." Due to factors that go beyond the customary rules at work in master-servant relationships, Wai and the writing child are unable to speak to each other: "When I call him to go still farther off and not to look again, he laughs so loud you can hear him all over the place. He laughs most of the time because he can't talk much English." The letter writer's makeshift assumption that the existing communication problem between him and his companion stemmed from Wai's inability to communicate in English, and *not* from his own inability to speak the language spoken by Wai and his people, marks the often overlooked, racially normalized hierarchy in child-child relations at the heart of this companionship. While the white child not only speaks but is also able to write about incidents in which language facilitated or impeded communication, the native boy's difficulty in speaking English is interpreted as utter unintelligibility. Even more striking is the white child's belief that Wai's inability to speak English automatically means that he cannot speak any human languages; while, rather proprietarily, the white child viewed himself as a proficient user and "owner" of a language (which happened to be the right one, too, he believed), he stripped the Native Hawaiian boy of *his* language, replacing his ability to speak (a sign of a particularly human form of communication) with his almost manic proclivity to laugh, thus suspending Wai's right to claim ownership over his mother tongue.

As young travelers staying in fashionable hotels, well-off American children often reduced the non-American world to a mere site of passing attraction—an undifferentiated space where they could consume the "foreign" while remaining American. An unsigned letter by An Admirer of St. Nicholas, printed in October 1890, provided an exhaustive account of such a trip to Mexico the previous winter. "I do not believe," the letter declared, "many of your readers realize that there is so strange a country so near our own. In Europe," the writer continued, "you cannot find cities any queerer or people any stranger than these." As if an ekphrastic account of an exotic fashion show, the letter wrote of a group of "picturesque" Mexicans who wore "broad-brimmed 'sombreros' and many-colored 'serapes'" and "blue 'rebozos' wound gracefully around their head and shoulders." The letter writer further reported that despite their fascinating and attractive appearance, the pair of leather sandals they bought in Mexico smelled so strongly that, upon returning home, they had decided to "keep them in the cellar."

Complimenting Mexico City on its broad streets, fine and numerous drugstores, and fantastic jewelry stores, the anonymous writer shared a story of an encounter with a young Mexican in a decidedly less celebratory tone: "The boy who answers your bell [in the Iturbide Hotel] is a true native; in vain you use your phrase-book, which never has in it the word you need; in vain you make frantic gestures, and at last you give up in despair." The letter writer had not, however, given up entirely. Again, as if a variation of the encounter with Wai Waiole in Hawaii, the letter writer seems to have believed that the communication problem resulted from the Mexican boy's inability to understand and follow orders and not his own faulty knowledge of Spanish: "After we had been there a while, he learned to understand a few words, and the next American must have had an easier time." Satisfied to have educated the Mexican boy, the letter suggested that any encounter between children of the two neighboring countries was in effect a tale of class and racial hierarchy, failed expectations, and small-scale but successful completion of a patiently executed civilizing mission in the shape of an encounter between (white) child and ("true native") child.

More often than not, rather than focus on the people they met, the letters recorded their trips to the non-American world exclusively in terms of a picnic that would take them far away from civilization and into the wild: the architectural wonders that children visited, the festivities they attended by chance, the scenery and the sites they got to explore, and, above all, what they did to the places and people they stumbled upon.[37] Whether recorded by Daisy and Victoria, siblings who had to leave home because their cousin Isabel was ill (*St. Nicholas*, November 1890); by Grace W. L., who sent a long letter to *St. Nicholas* (November 1890) to talk about her adventures in Far Eastern castles, coastlines, and mountains; by Harriet B. S., whose letter to *St. Nicholas* (printed in May 1890) related a story about Antwerp's "very funny" tradespeople; or by Edith M. K., daughter of a missionary who was revising the Hindi Bible in Dehra Dun, India (*St. Nicholas*, May 1894)—the world, in the eyes of these children, consisted of a litany of personally captioned images and variously scaled mental maps.

Breaking down the temporally ordered frame of going far only *after* having grown up (which I discussed in chapter 1) as the envisioned, judiciously sequenced task that children were assigned as future stewards of the emerging overseas U.S. empire, American children abroad would grow up *exactly* at the same time that they went far. This was particularly true for American children who lived abroad, often at boarding schools and in the company of children from all over the world. These children had their American nationality attached to their home front when living at a missionary station in Persia, at a boarding school in Switzerland, at a diplomatic outpost in Tangier, or when aboard ships that carried the U.S. flag across the oceans. The term "home front" has most commonly been invoked by historians and Americanists in examining the lived experiences of World War I and World War II and in scholarship that has recorded the transformative force of these two wars on the local and national levels in the United States, including on women and children who did not go to war and yet were profoundly affected by it.[38]

However, it can also be applied with a modified meaning in the turn-of-the-century United States, especially with regard to the so-called civilizing and Christianizing missionary activities of Americans abroad. In this context, the home front brings together notions of belonging, familiarity, safety, and comfort at both the edge of wilderness and the edge of war.[39]

Aside from ongoing spatial tensions at work across the national expanse of the American home, the notion of the "home front" marks the borders of the metaphorical imaginary of a portable collective U.S. home as it was carried around the world. It did so in two broad ways, marking the two-way borders of the nation: first, for immigrants who hoped to leave the quarantine stations and the checkpoints behind in order to enter the country, and second, for Americans, including children, who planned to leave the United States for short and long stays abroad. Distancing themselves from the inhabitants of the non-American world, these children made sense of the borders of their purportedly "exceptional" home front as a "frontier": a thin spatial shield between U.S. civilization and whatever they believed was standing in contrast to it, be it a lack of self-discipline, an inability to communicate in English, indolence, poverty, and, above all, a stagnant or waning state of civilization.[40] Indeed, as noted before, not a single letter in the period under study gave an account of encounters between Americans and people whom they viewed as their equals.

A wide number of letters printed in *St. Nicholas* and *Harper's Young People* divulged their writers' deep-seated sense of unhomeliness among non-Americans, nonwhite populations, or non-Christians, while others talked about the magazines as their only friends—tangible bits of America that reminded them of a sorely missed home. In a short letter, eleven-year-old Harriet B. S. (*St. Nicholas*, May 1890) told her readers how lonely she felt while staying in Antwerp, Belgium: "ST. NICHOLAS seems like a dear old friend, as it follows me across the ocean." Writing in a comparable tone, Thomas Harper Goodspeed, also eleven years old, wrote about his life at the Collège Gaillard in Lausanne, Switzerland (*St. Nicholas*, September 1898). Taking his readers on a quick tour of the surrounding Alps, Thomas reported that he was the only English-speaking boy among the pupils attending the school, who were from Africa, Chile, Brazil, France, and Switzerland. After listing all the boys at school, he gave an account of his loneliness as an American abroad. Resistant to the idea of calling Lausanne home, Thomas wrote: "When I am in a foreign land my greatest pleasure is to receive the ST. NICHOLAS." Overall, the very act of sending letters back home underlined English as the linguistic marker of Americanness and worked as a welcome reminder to these young

Americans abroad that they were still members of a spread out and yet unified imagined community—the home front—that was not only American but also young, literate, and wealthy enough to afford overseas adventure and education. Parents who worried that their children might grow up too foreign to their American roots particularly welcomed letter writing is an act of community building. In fact, by writing letters to American children back in the United States, adults believed that their children would reconnect to the United States. As residents of American homes abroad—spaces that, in their own way, could be considered domestic to the United States in a foreign sense—these children reported that even as they kept exotic pets, had foreign friends, and spent time with only a handful of English-speaking children in the neighborhood, they tried to remain true to their American roots. In this vein, many children complained in their letters about spending a boring Fourth of July in whichever corner of the world they were living: "As I was in Austria on the 4th of July this year," Robinson N. wrote to *St. Nicholas* (printed December 1891), "I was very much disappointed in not having the lovely fireworks that we see at home." Similarly, A. A.'s letter (printed in *St. Nicholas* in April 1895) mentioned that while he and his brother, students "in exile" in Dresden, Germany, observed "all American holidays most faithfully," they had little chance to celebrate the Fourth of July, "for there is no noise allowed here."

A large number of letters sent by American children abroad involved outbursts of patriotism and expressions of nostalgia for home. Expressing awe at how magnificent and beautiful they found Paris, they would quickly conclude—possibly out of guilt for having enjoyed their trip abroad, or as a result of a deep-seated sense of national pride—that it was nothing in comparison to the United States. Helen McC. wrote one such letter to *St. Nicholas* (printed November 1891). After she gave an account of her and her two sisters' life at a boarding school near Paris, Helen rejoiced in the fact that they "have a great many friends here." And yet she confided in her readers that "we wish to go back to our native land." Harboring similar sentiments, Cyril Cecil S. concluded in his brief letter about a family trip to Rome: "I like Europe very much, but America is the place for me." In a letter printed in *St. Nicholas* in November 1894, and writing from Poros, Greece, young Gardner R. echoed Cyril as he finished his letter with "I am homesick for America.... I like it here, but I like my home better." Other children confessed rather profusely how they wished to go back to the United States after coping with troubling or peculiar experiences outside the United States. At the end of a neatly composed letter, An Admirer of St. Nicholas admitted that "to see the horses gored, the bull killed and dragged out, is an experience I never wish to repeat again." Watching the bullfight, the writer opined, "you get a good idea of the Mexican." Reporting the frenzy and the unpleasant aftertaste of witnessing the bullfight, during which the writer misattributed bullfighting to Mexicans (and not their Spanish colonizers, which itself had ancient roots in prehistoric Mediterranean, Roman, and Greek cultures of sacrifice), the letter ended with an overly mawkish outburst of patriotism: "I confess I never felt more like singing 'America' than I did when we crossed the Rio Grande and were again on native soil." Prompted, at least partly, by a combination of home*less*ness, disorientation, and out-of-place-ness that one experiences away from home, children's deep sense of homesickness and the desire they felt to return home take us back to yet other home geography questions: Did American children look for but fail to find other homes, or would-be homes, outside the United States? And if so, was this failure and the ensuing anguish an outcome of children's inability to read maps and to navigate spaces in order to spot other homes—that is, scraps of "civilization" scattered across the globe? Or was their homesickness the result of their inability to imagine other possible drafts of home unless they followed an identical spatial order and outlook as their homes back in the United States and unless their inhabitants looked and sounded familiar? In other words, were children inadequate practitioners of home geography, or was home geography an inadequate approach to encountering the world? We can further ask, in retrospect, whether all American children felt so miserable when they were abroad, or whether only those letters found their way into The Letter-Box and Our Post Office Box that expressed patriotism, homesickness, and a sentimental desire to return to the United States.

During the final years of the decade, and especially with the outbreak of the war between the United States and the Spanish Empire in the latter's colonies in the Caribbean and, later, in Southeast Asia, a few notable letters took patriotism out of the realm of sentimentality and into politics. In lengthy, comparatively well-written letters, children remarked on how their lives as Americans abroad had changed in the wake of the ongoing war. These post-1898 letters had a heavily patriotic tone and exposed their writers' penchant for taking sides against anyone who held opposing political views. During the time the war had spread to the Philippines and with the turn of events that broke the alliance between Americans and the Filipino army against the Spanish, fourteen-year-old Elizabeth Haviland Brown wrote a letter to *St. Nicholas* (printed October 1898) from Weimar, Germany (figure 4.2). Elizabeth prided herself on her pronounced sense of patriotism, as a result of which she had

> WEIMAR, GERMANY.
>
> DEAR ST. NICHOLAS: I am a little American girl of fourteen years. We have been living in Weimar for little over a year, but expect to return to America in the fall.
>
> I often am very angry with the Germans, as some of them sometimes make fun of our flag (we have one in our garden), or they'll make some mean remark about America. When I say this I do not mean it of all the Weimarians, because some of them are very nice indeed.
>
> I have a little canary bird, which I call "Dewey."
>
> My younger brother goes to school here, and one day a boy said to him: "*Amerika hat keinen Verlust.*" The last word he did not know; but, as the boys sometimes make fun of America, he thought the boy meant our soldiers had no courage, and therefore gave the boy a beating—not a very bad one, as he (my brother) was much the smaller. After he came home he told us about it, and we had a good laugh, as the boy meant, "America had no loss"—referring to the battle of Manila.
>
> They have here, three times a year, a *Jahrmarkt*, or yearly market. At those times they sell almost everything—all sorts of clothing, fruits, vegetables, candy, meat, fowl, porcelain, earthenware, toys, etc. They are quite interesting, though the only things we care to buy are curious pieces of porcelain and earthenware.
>
> We all enjoy the ST. NICHOLAS immensely. The stories are fine, also the illustrations. I think Birch's are beautiful. My sister and myself often copy them. We have subscribed for the ST. NICHOLAS for many, many years.
>
> With much love, I am ever your little friend,
> ELIZABETH HAVILAND BROWN.

FIGURE 4.2 Letter by Elizabeth Haviland Brown, *St. Nicholas*, October 1898 (courtesy of HathiTrust).

called her pet canary Dewey—most probably to honor Admiral George Dewey (1837–1917), the "hero" of the Battle of Manila. Frustrated with how their German neighbors viewed the ongoing war, she confided in her readers: "I often am very angry with the Germans as some of them sometimes make fun of our flag (we have one in our garden), or they'll make some mean remark about America."

Elizabeth continued the letter by recounting an unpleasant event that had happened to her brother at school: "One day a boy said to him: '*Amerika hat keinen Verlust.*' The last word he did not know; but, as the boys sometimes make fun of America, he thought the boy meant our soldiers had no courage, and therefore gave the boy a beating—not a very bad one." She then observed in a nonchalant tone that the family had "had a good laugh" after hearing what had happened once they realized that the German boy had actually meant "'America had no loss'—referring to the battle of Manila." Given the letter's date, October 1898, Elizabeth was almost certainly referring to the U.S. victory under Admiral Dewey in the so-called Mock Battle of Manila. By fighting a mock war with the Spanish forces in order to deceive the Philippine Revolutionary Army, the Mock Battle of Manila had facilitated the surrender of the city to the American forces. Having been staged following days of covert negotiations between the two imperial armies (the Spanish and the American), the battle was fought on August 13, 1898. This mock battle is believed to have put an end to the Spanish-American War, resulting in the fall of Manila into American hands, while also marking the start of the Philippine-American War.[41]

Non-American children, too, reacted to the war, and the magazines printed their letters. Geoffrey Dale sent his letter to *St. Nicholas* (printed September 1898) from Ashfield in Sydney, Australia. Born to an English family, Geoffrey dedicated almost the entirety of his letter to praising, and praying for, Americans soldiers, confessing himself to be "very anxious for the fear the war should prevent ST. NICHOLAS from coming here." Geoffrey finished his letter with a personalized complimentary closing: "Sincerely hoping you will win in the war, and that ST. NICHOLAS won't be taken captive or killed, I remain your lifelong admirer." On the other hand, and perhaps to remain neutral toward these non-American perspectives, the same issue contained a letter from French sisters Odette, Vivianne, and Marie Antoinette M. De. B. that maintained the exact opposite: "We like very much the ST. NICHOLAS—in fact, it is the only thing we love in America, because," they retorted in a stern tone, "all our sympathies are for the poor Spaniards. We have friends in the Spanish army."

Examining children's letters as a whole sheds light on the ways their young writers placed themselves in the spatial expanse of the United States as entitled members of family units and as residents of the national household at the same time that they earmarked foreigners as deracinated individuals who did not have any rights over their names or clothes or over where and for how long they had to wait before entering the United States. In their overall brevity, children's letters further create a registry of an endless number of moments at which children—suffering the hardships of travel and experiencing homesickness while also enjoying the freedoms that being in a foreign environment and away from home offered—stood at the threshold of a larger world, felt disappointed at the sight of famous religious sites in the Holy

Land, or found the borders of "home" and "non-home" blurred when visiting their immigrant parents' hometown back in Europe.

PUBLICLY ACCESSIBLE AS FAN LETTERS, children's letters to periodicals have a life of their own. They are labor intensive yet freely available cultural goods that a rising generation of literate, loquacious "writing children" produced in and about the homes or home fronts where they resided. They are written at the invitation and under the supervision of adults, parents, teachers, governesses, and serve—or fail to serve—their child and adult addressees' expectations. And by entering the pages of a children's periodical, they continue their life as documents that may or may not find a place in the historical archives. Regardless of their length, these letters present us with striking evidence of numerous, evolving drafts of their writers' tentative cognitive maps in racial and commercial terms. Perceiving the world in its totality as a catalog of interrelated places to explore as tourists-observers or as young civilizers– cultural critics, a great many literate American children viewed their encounters with the world in ambivalent terms. Indeed, the contradictions between the unprecedentedly global lens through which some children observed their home as part of an interconnected web of homes and the curiously local optic with which others looked at the United States as a world in and of itself further reflect the complexities of a society that, although geared toward a new phase in its economy and a new chapter in its foreign policy, continued to accommodate locally oriented mindsets that regarded the nation as a self-sufficient entity—inward looking and far from imperial.

Constitutive of a sort of cognitive malapropism, these public letters divulge a great deal as to what their young authors noticed, ignored, internalized, remembered, imagined, reconstructed, felt, or took back home with them in their encounters with the world, in their avid perusal of juvenile periodicals and travel books, and through their innermost daydreams of adventure in the form of sketchy maps that they then translated into geographical puzzles and short fan letters. Indeed, indicative of the unpredictable, provisional, and creative features of cognitive mapping, and symptomatic of a high degree of intergenerational continuity, children's letters marked the loose outer borders of the open-ended, neological realms that their composers developed over the years in their geographical imaginings of the world. It is of course possible that in our encounters with these letters, we do not readily notice the various ways that these neologisms remain faithful to or deviate from cartographic practices and geographic narratives intended for children by adults.[42] It is, however, certain that such youthful imaginings did not fully discard adults' intended "projects" for the nation's future standing in the world. Even so, children's attempts at altering and assigning new meanings to this very world and envisioning it in ways that deviated from the principles that turn-of-the-century adult Americans had established attest to children's awareness of the power of the platform from which they shared with others glimpses of their ever-changing cognitive maps of the world, in epistolary form.

In calling for a more methodical examination of children's artifacts in this volume, I concede that at the same time that we make sense of children's writing as challenging and yet inviting, we also run the risk not only of cementing the age-old child–adult binary more than it already is, but also of forgetting the dynamic afterlives of children's writings when read by their original authors' children or grandchildren. Thus, while laboring to spot and then explore repositories of childhood, we need to keep fresh in our minds that the tensions between childhood and adulthood, on the one hand, and between childhood as an adult-bred artifact versus childhood as a lived set of exposures, experiences, and expressions by generations of flesh-and-blood children, on the other hand, merge to form a permanent ideological, material, and temporal slippage. This slippage, in turn, sets nation- and empire-building projects designed by adults in competition with the lack of emotional and cognitive investment that frequently characterizes children's attitudes toward the cause of nation and the cause of empire. In the case of the child-adult interdependencies studied in the present volume, for example, while adults' attention was often fixated on the future and as they labored to design paths that ensured the impending and much-desired expansion of the nation and the empire, children were often busy grappling with the present and with what it meant to be constantly in flux, growing up in body and mind, and growing out of childhood. I revisit this tripartite slippage in the conclusion.

Conclusion
Huckleberry Finn in the World

As if pushed to board a world-bound balloon, the turn-of-the-century, globally emerging American empire had come a long way in projecting the territorial expanse of the nation onto the world and in reinscribing its frontiers beyond its erstwhile continental coordinates and on unprecedented scales. Indeed, in the material under study in *Citizens and Rulers of the World*, the inevitable fate of the child, bound to either learn to grow up into adulthood or perish, is intricately woven into the projected fate of the United States at the turn of the twentieth century to either rise to overseas global empire or to eternally grieve the closing of its western frontier and the suspension of its expansionist national identity. With close textual and visual readings of the marginalized and often forgotten, yet heavily curated and edited accounts of childhood in the 1890s United States—what I understand as an example of the multipotent microhistorical frame of analysis that I call "the semiotics of the overlooked"—I have focused my arguments in this book on Americans' efforts around the turn of the twentieth century (and beyond) to assume an increasingly outward-looking national and imperial identity in cultural geographical terms. This entailed attempts to come to terms with a conflation of relations as citizens of a comparatively young nation and a spatially unsettled empire. As I have argued, while the ways Americans carried themselves as the purported citizens and rulers of the world changed, they further grappled with a series of contradictions contained in the concept of home geography— that is, the idea that the world consists of a network of variously distanced, comparable (but also at times categorically incomparable) homes/localities.

As I maintain in the preceding chapters, turn-of-the-century encounters of Americans with the world—whether material, perceptual, or representational—were experimented with and textually and visually expressed in new editions of geography primers, dissected maps, geographical board games, and juvenile magazines, as well as in the specific language that adults and children deployed for intergenerational exchange. Turn-of-the-century American adults added to the already massive body of world maps, revised them to fit the nation's expansionist geopolitical and economic priorities at the dawn of a new century, and modernized geography by defining it in terms of the complex reciprocities between human beings and the environment.[1] In response, a great majority of privileged American children were given the opportunity to write about their trips abroad, comment on their tours of prisons and quarantine stations, and turn their encounters with non-American peoples and places into puzzles. American children took their knowledge of world geography out of schoolbooks and appropriated it in strikingly diverse and subversive ways. As letter writers and composers of puzzles (but also as diarists, essayists, and poets, among others), as students and playmates, they authored a world that was not only intimately out of scale and variously impressionistic but also categorically unplottable. They shared scraps of their tentative cognitive world maps in the letters and geographical puzzles that they sent to their favorite magazines. Throughout, the threads holding the chapters together weave adults' means of producing world geography knowledge into children's often-slippery patterns of receiving, appraising, and appropriating it.

In order to round off the study of the links between artifacts of childhood and their geopolitical contours when offered to, consumed by, and further reproduced or refused by children, I start this concluding chapter by briefly examining the enthusiasm with which the juvenile periodical press both celebrated and promoted the production of geographic knowledge for the world beyond the national borders that the United States was formalizing at the beginning of the new century. I then focus on the resultant proliferation of schooling and teaching methods, policies, and practices in the Philippines at the start of the twentieth century in order to answer the twofold question: Did the proliferation of material (a) to teach children in the Philippines a U.S.-authored version of global geography and (b) to teach American children to include far-flung parts of the world like the Philippines as "belonging to them" convince children to assiduously engage in the material processes of sustaining and celebrating the American empire? This section entails observations on some of the ways that children—in their varied roles as readers, learners, players, writers, puzzlers, world travelers, and even cartographers—engaged with various geographical aspects of their national identity. The discussion of imperial knowledge production and regimes of imperial knowledge dissemination is rounded off with concluding remarks on the most typical image of Twain's Huck Finn: smoking a pipe, purportedly uninterested in and uninformed about his country's global imperial ambitions and schemes.

Huck in Baghdad: A Call for Rewriting Geography Books of Empire

Toward the end of the nineteenth century, and thanks to the incontrovertibly more central place the United States now occupied in the constricted circle that the "family of nations" consisted of, the knowledge and the imperialist aspirations that the periodical press shared with young readers were shaped under the aegis of forceful geopolitical and commercial interests that the United States had in the world at large. Laced with patriotism, envisioned on scales both smaller and larger than the nation, and keenly exported by Americans to the world, U.S. geographic knowledge production had finally matured. In the following few pages I will do a close read of a short story that appeared in the juvenile periodical press as an example of this maturation that scripted the nation's growing yet firm role as not only a consumer but also a producer of geography.

"The Persian Columbus: An Oriental Fantasy" appeared in *St. Nicholas* in 1892. The story is set in summertime Baghdad, and the year is 870 in the Islamic calendar. The story starts with a deeply troubled Caliph Haroun Al Huck-El-Berri in his palace. Holding an orange in his hand, he wonders: Is the world flat, as "they think in the Orient"? Or is it, as "Christoval Colon," the Western mariner from Genoa, has claimed, round? Unable to answer the age-old question by himself, the fat, old long-bearded caliph—namesake of the young protagonist of Mark Twain's novel *Adventures of Huckleberry Finn*—summons his shrewd vizier, the Seven Sages of Bagdad, and the commissioner of public schools, demanding that they give a satisfactory answer to the question in fifteen minutes. They all fail (figure C.1). Flabbergasted as he is by their failure to answer such a foundational geographical question, the story goes, the caliph takes it upon himself to travel to the end of the world in order to find out whether or not it is like the orange he has in his hand: "If I begin here [pointing to the orange with his index finger] and move onward, my finger soon passes completely around the orange and returns to the point whence it started." Having established this, the caliph embarks on a journey to explore the world with a royal army.

In the middle of the journey toward the North Star, the caliph falls asleep. Afraid of falling off the edge of the flat earth, and in a state of confusion caused by horse tracks on the sand, the caliph's vizier makes the royal caravan take a gradual U-turn back toward Baghdad. As the caravan is about to reenter the city, Caliph Huck wakes up from his sleep, which is not be disturbed by

160 Conclusion

"TELL ME, YE IGNORAMUSES," SAID THE CALIPH, "IS THE WORLD ROUND OR FLAT?"

FIGURE C.1 Haroun Al Huck-El-Berri and the Seven Sages of Bagdad, *St. Nicholas*, December 1892 (courtesy of HathiTrust).

"a little thing like riding around the world." Upon seeing Baghdad, the caliph excitedly announces that "the world is round ... [and] that Persia runs completely around it in one direction, and pretty nearly around it in the other." Reminiscent of President McKinley's presidential order to the cartographers at the War Department to "put the Philippines on the map of the United States," the story ends as the "discovery" leads to the caliph's royal order that substantial changes be made in all geography books that have been published across the empire, modifying their false conceptions of the world, its shape and size, and where in its expanse the empire stands.[2]

Written by Jack Bennett (1865–1956), "The Persian Columbus" forges bonds between the future-oriented course of growing up and the forward-looking cause of going far in a tone comparable to Fred Cartmell's trade game in the lesson/story from the *Picturesque Geographical Reader* examined in chapter 1. What is more, Bennett's story rather playfully engages with American adults' views about the globalizing contours of the U.S. empire, as it also parodically parallels the state of world geography knowledge in the United States with a number of fictional events in a Middle Eastern court. Joining the public festivities to commemorate the four hundred year anniversary of Columbus's landfall in the Americas, the short story seems to have adopted the general plot of Twain's 1889 novel *A Connecticut Yankee in King Arthur's Court*.[3] In Twain's novel, a New Englander called Hank Morgan is transported back in time to the court of King Arthur in the sixth century A.D., where he survives by fooling medieval English people with the help of modern science and technological advances. Though loosely inspired by the novel's time travel, Bennett's "The Persian Columbus" could also be read as a prophecy that foresees coming changes in the role the United States was going to play in the world, and the reversal in its position with regard to the material that defined and taught world geography at the turn of the century—as a global producer, rather than a mere parochial consumer. Comparable to Twain's arguments in *Tom Sawyer Abroad*, "The Persian Columbus" acknowledges the clashing "geopolitical imaginaries" of its various characters.[4] Furthermore, the story reminds its young readers that any imagined community—in its Andersonian sense—though assumed to be homogeneous, in fact enjoys a conglomerate of competing imaginaries. Any community's members—young or old, orphan or emperor, the story further insists—follows unique sets of spatial practices in order to make sense of the expanse of the world and to find their way in and around it.

Although a short story intended for children, "The Persian Columbus" alludes to several central and timely political matters that had surfaced in the United States around the time it was published. First, the story hints at the transitional state of geography as an academic discipline and as a school subject during the final decades of the nineteenth century and reflects the concerns of professional geographers and schoolteachers about the new definitions of geography and the rewriting of geography textbooks as a national agenda. In the story, even the commissioner of public schools, who, as the official adult body behind large-scale changes in the contents and pedagogic methods in (American) public schools is supposed to have a certain understanding of geographical facts to begin with, struggles to answer the urgent, basic geographical questions of the time. Second, the adventurous Huck, youthful despite his age in this short story, explores the world as an imperial adventure—interestedly yet indolently, as the original Huck would. As an "Oriental" caliph, he finds out that the notion of living on a flat earth is nothing but a myth; that the Western Christoval Colon had been right in insisting, and had proven (through the so-called discovery of the Americas) that the world is round; that, as it takes the caliph only three hours on camel back to travel around it, the world is small; and, finally, that it is contiguously imperial, with Persia—the pioneer worldwide empire of the ancient times—right at its center.

By renaming the Abbasid caliph Haroun Al-Rashid (A.D. 764–809) as Haroun Al Huck-El-Berri, Bennett attempts either to rejuvenate the grownup caliph in the form of an already familiar American juvenile adventurer,

Huckleberry Finn, or to age the young American hero into a fat old caliph who changes the understanding of world geography in the entire expanse of the Persian Empire. Either way, the story's young reader encounters a keen adventurer who is both proximate (Twain's familiar and popular teenage Huck) and distant (garbed as an emperor and situated in imperial Persia of the fifteenth century—which, ironically, coincides with Columbus's landfall in the Americas). Here, stereotypes converge to revise geographic knowledge: the American boy, adventurous and imperial, is also the naive "Oriental" caliph who—thanks to Christopher Columbus and his imperial sponsors in the Spanish court—"rescues" his people from false beliefs about the world. Though garbed as an emperor of the "Orient," Huck remains unmistakably American for the young readers of the story: adventurous, inquisitive, playful, and upright. In effect, he symbolizes the qualities American adults hoped their children would acquire as young yet maturing "stewards of empire."

Excited as Caliph Huck is to personally verify the three-dimensional nature of planet Earth, he is far too lazy to stay awake during the journey. Furthermore, the caliph naively trusts his vizier, who—based on the firmly held belief that the world is flat—tries to fool him. As such, Caliph Huck is "American" to the extent that, once he is confronted with Western facts, he singlehandedly manages to save the infantile "Orientals" from false geographic beliefs—thus fulfilling the role of a genuine Western "civilizer." On the other hand, he is convincingly young and naive to the extent that he has little understanding of the world's shape and size and, yet, is deeply curious to know about it.[5] If, therefore, we accept the caliph as the young American Huck, then his conclusion about the size of the world and the place of the Persian Empire at its center could be directly translated into corresponding beliefs about the U.S. empire. The story conveys the message to its young readers that the world is a contiguously imperial, imagined community centered around the United States and that the American traveler, as a citizen of this globally mappable center, can navigate it.[6]

Finally, and most importantly, the story serves as an endorsement of the common justificatory argument of the time that it is empires' right as well as their responsibility to produce and disseminate (geographic) knowledge across the globe and to their subjects both at home and in the so-called peripheries. After all, the history of modern empires—multigenerational projects of domination and control—has also always been a history of didactic imposition. In concert with the self-professed right and responsibility to disseminate knowledge that is reflective of the empire's supposed moral and scientific superiority, such knowledge was produced, the story seems to convey, based on the empire's scientific breakthroughs and on its purportedly authoritative position as civilizer and educator of the colonized.[7] As the following discussion establishes, imperialist Americans wished to believe that it was *their* responsibility as well as *their* right to educate their new "subjects" and to overwrite the allegedly incomplete and erroneous forms of knowledge that were offhandedly earmarked as inferior only because they were labeled as "Indigenous." After all, "The Persian Columbus" was published at a time when the slow but sure "maturation of the United States" in the image of empire and the ensuing set of rights and responsibilities it was going to assign itself by the dawn of the new century seemed inevitable, violence toward some and welcomed by others.

Huck in Manila: Imperial Right and Dissemination of Knowledge

In the same spirit as the prophecy put forward by Bennett's story as a reflection of public sentiments about the changing role of the United States in the world, and right at the turn of the century, when the U.S. government began to take the full measure of its newly won colonial possessions in the Pacific and the Caribbean, Americans began to devise strategic plans to address the question of educating the empire's new subjects. While Bennett's short story addresses American children, instructing them as to how to perceive the empire and their responsibility as part of it to right geographical wrongs on a global scale, the geography schoolbooks sent to the Philippines instructed the colonized children (and adults who were treated like children) in the ways they were supposed to understand the American empire and their place in it. When U.S.-produced geography textbooks and world maps reached the Philippines, it was hoped they would help stretch the "national" beyond its contested yet contiguous borders on the North American continent in order to make room for those that were now a part of its global imperial expanse. Decades of anticolonial struggle and the resulting severing of ties between the Philippines and the United States stands witness to their resounding—if deliberate (from the perspective of white supremacists) and (made) inevitable (by anticolonial forces in the Philippines and elsewhere)—failure.

Starting at the primary school level, the plan to educate Filipinos in English and in the image of the American child was formally proposed at the proceedings of the United States Philippine Commission in Boac, Island of Marinduque, on March 15, 1901. The commission's report suggests that its members were unanimous in their endorsement of the plans for English teachers and new school supplies to be sent to the archipelago.[8] Announcing that around a thousand American teachers were to arrive in the Philippines to train Filipino teachers and to teach Filipino students, the report declared optimistically: "Our most satisfactory ground for hope of success in our whole work is in the eagerness with which the Philippine people, even the humblest, seek for education."[9]

In the 1901 report by the Philippine Commission on the operationalization of U.S. imperial rule over the archipelago's public health, finances, army, justice system, and so on, the longest single report was indeed that of the general superintendent of public instruction.[10] Deciding on the course the pedagogical encounters between the colonizer and the colonized were to take in the years to come, the report included extensive information about the current state of public education on various islands and provinces of the Philippines as well as exhaustive instructions about school subjects, "patriotic" exhibitions, monthly schedules, and so on. Those American secondary school and college graduates who wished to seize the opportunity to serve as teachers in the Philippines had to formally apply to and pass a number of exams, including two exams in physical and political geography.[11] The political geography exam was referred to as an "Examination in Current Topics" and included questions on the "Boxer Movement," the "treaty of peace between the United States and Spain," and several questions on current relations between the United States, Cuba, Puerto Rico, Hawaii, and the Philippines.[12]

In the section "Supplies Received since January 1, 1901," Fred W. Atkinson, the general superintendent, complained that while eager learners, Filipino children "are not particularly careful with schoolbooks, and in the rainy season the books suffer very much from getting wet."[13] However, despite the damage to school supplies such as slates, pens, and chalk that occurred during their transoceanic transport, and while the exasperation with Filipino children's behavior (and the weather!) cast doubt on whether additional American schoolbooks should be supplied, the report suggested that Americans were in fact considering whether they should send even more books to the Philippines.[14] Since January 1901, the report maintained, the commission had received $133,510.00 worth of schoolbooks (including the price of the books and the transportation costs) and $37,271.48 worth of school supplies.[15] Among these books were geography texts, including a staggering thirty-five thousand copies of *Frye's Geography*, ten thousand of *Tarbell's Geography*, ten thousand of *Carpenter's Geographical Reader, Asia*, ten thousand of *Big People and Little People of Other Lands*, and ten thousand of *Guyot's Geographical Reader, North America*.[16] The commission had also received supplies such as

1,139 U.S. flags, 500 8-inch globes, 100 maps of the Pacific Ocean, and 500 maps of the world and the United States, while a request had been made for 1,000 more U.S. and world maps.[17]

Within a matter of years, in addition to the school primers that were originally compiled for educating American children back home, and existing volumes with extensive appendices on the Philippines, newly compiled volumes were sent to the islands. Prescott Ford Jernegan (1866–1942), for example, compiled volumes on civics, history, and geography for Filipino schoolchildren.[18] His *Philippine Geography Primer* opened with a map of the world, "showing the United States and its possessions."[19] Including a hundred and fifty short lessons in geography, the primer was designed for a full school year of instruction in the third grade.[20] Subscribing to home geography as its prime teaching method, the volume focused mainly on the political and commercial geography of the Philippines and its relation to the United States. By enlisting young children in U.S. empire's colonial operations and successfully disguising exploitation as education, Jernegan asserted in the preface that *Philippine Geography Primer*'s most urgent objective was to inform Filipino pupils about the state of their country's natural resources and its export products upon its transition from an ailing pair of imperial hands—that is, the Spanish Empire—to a more adept pair. Jernegan then sketched a meticulous plan for a future of prosperity, a plan through which "the resources of the land and the people may be *more fully utilized* [emphasis in original]."[21]

The general argument for the necessity of compiling and adopting new volumes was that they would meet the needs of the inhabitants of the islands and prepare them for their future civic roles as subjects, though not citizens, of the U.S. empire more fully than the primers that had formerly been published exclusively for American children. For instance, right after the preface, Jernegan found it necessary to dedicate a few pages of his primer to pedagogic recommendations to Filipino teachers: "Do not let the children recite like *parrots*. Make them think. . . . Be sure that you *find out the reason* and learn how to *explain* it. . . . Always ask yourself and the children of what *use* the knowledge gained in a lesson is. Ask how it may help them to become good *farmers* and *workmen* and *citizens*. . . . *This is why we study geography* [all emphases in original]."[22] Despite numerous efforts to write the Philippines into the fabric of the U.S. global presence, Jernegan sustained a bizarrely patronizing tone throughout the book. At one point, he blamed Filipinos for their "wasteful and foolish" behavior in burning trees, natural resources that "will be worth a great deal of money in the future."[23] Later, directly commenting on Philippine geography, he made sure to explain that "Manila . . . is the only city of the Philippines that is known all over the world."[24] Summarizing the city's colonial heritage in the section on history, Jernegan was swift to conclude that under American rule, Manila "is now one of the best-governed cities in the world, and one of the finest cities in the Orient.... *Every Filipino should visit Manila* [emphasis in original]."[25] When it came to the geography of the North American continent, Jernegan made frequent comparisons between the United States and the Philippines, their size, wealth, population, climate, political systems, natural resources, and industrial products. These comparisons later served as the basis on which he justified dependency of the latter on the former in governance, the passage of laws, and education.[26] Avoiding writing in the language of empire and yet closely resembling it, Jernegan enumerated the global "Possessions of the United States," laboring to draw a glorifying and meticulous picture of the various islands and regions that the United States "possessed." The section went to great lengths to liken Filipinos to Puerto Ricans, Samoans, and Hawaiians in race, industries, and natural resources, as if it were only "natural" that they belonged, in harmony and homogeny, to the margins of the great young empire.[27]

Huck at Home: Concluding Remarks

Despite American adults' devoted and systematic interest in imbuing children with patriotism and in preparing them to assume critical roles in sustaining the U.S. empire and fulfilling dreams of their parents' generation as to the future of the nation and the empire, the discussions in the preceding chapters have made it evident that children consumed geography playfully. Indeed, reminiscent of Huck Finn lying on his back "for a nap before breakfast," the expansionist scripts of the nation were not always a priority to them, nor were the nation's political concerns beyond its borders always in their hearts.[28] As young citizens of a rising global empire engaged in reading, writing, and playing, American children were, in Tara Zahra's words, "nationally disinterested."[29] They could be passionate patriots and yet invest their attention in matters other than geopolitics. If, as I argued at the end of chapter 4, we are faced with an ideological, material, and temporal slippage between what adults present and what children preserve, then the questions become: Was American adults' investment in children as the future of the nation and of the empire pointless? Should childhood studies scholars bother to study children's magazines, playthings, and school primers? Should historians look at children's archives and examine world cartography in the province of the child?

The short answer to these pressing questions is in the affirmative. Studying children's archives reveals both traces of extensive adult investment in children's (geographical) education and socialization and the unique but purportedly ephemeral acts of comprehension and expression, compliance and deviation performed by children themselves. On the one hand, geographical games, periodical publications, and teaching tools, which mixed or else clearly bordered both didactics and entertainment for the first generation of Americans to reach adulthood by 1900, were sites of negotiating national identity—sites where spatial positioning, cartographic imagination, diplomacy and commerce, and questions of nationalism and imperialism all converged. Moreover, the indifference toward and the inaccurate invocations of the global in the case of children's geographical puzzles, their letters, or the doodles in the margins of their geography primers must be understood as complementary to and yet compensated for by the competitive, fact-based nature of the magazine puzzles that needed to be answered correctly within three weeks—not in order to achieve something as life changing as passing the entrance examination to Harvard University but simply to ensure that a child's name appeared in their favorite magazine.

To answer this set of questions in relation to the types of material examined in *Citizens and Rulers of the World*, on the other hand, it is evident that we all—child and adult—have elusive yet supposedly unswerving and highly individual mental images of what we come to know as the planet and its socio-spatial representations as the globe, at times referring to everything residing outside the discernible borders of our corporeal physique as "the world." Knowledge about any such external worlds—in their inaccessible totality, and with varying degrees of comparability to what any other member of human society refers to when deploying the same term—is not and cannot be pointed at easily, sustained inviolately, or analyzed objectively. To begin with, such knowledge is subjective; starting at the level of our very bodies, it is both experiential and volatile. The human body—the microcosm, the irreducible site in which power invests itself—moves through, experiences, and exercises its rule over (and its power is in turn delimited by) this socio-spatial "reality." In addition, this reality holds on to social discipline, ethics, and morals, and is inevitably a site of the microcosm's interaction with other micro- and macrocosms. Therefore, regardless of age, gender, class, race, sexual orientation, ability, and nationality, socio-spatial realities of the world are not a mere extension of individuals' imagination; far more significantly, these imbricated realities also function as sites where the relational, pluralistic, nonbinary exchanges between micro and macro, individual and collective, national and imperial, local and global, give shape to spatial imagination and knowledge.[30] At the same time, in microlevel cultural geography, "The world is . . . the sum of human experiences through their encounters with 'external reality,' which cannot be accessed other than through the human mind."[31] Therefore, getting to know the world is, as I noted before, a tentative project of cartography in progress: a subjective, unsettled, bodily, and shifting cognitive—or at least imaginative—exercise.

In light of this, at issue in this volume is to tap into children's archives in a data-driven sense—that is, as a means of tallying how many children knew the place names that the letter X stood for and how to correctly spell those toponyms, or even how consistently children subscribed to narratives of American imperial superiority—but also to remind ourselves of the variation and deviation in perception that is bound to emerge from interacting with the same set of texts, tools, and toys. Variations and deviations in perception result from the fact that depending on where we gain geographic knowledge as well as on our backgrounds and intentions, our knowledge as humans is conditional. Geographic knowledge depends, among other factors, on whether individuals gain this knowledge through their own active (con)quest and (un)intended life encounters with different peoples and parts of the world, or whether they are fed a body of information, interpretations, and statements within a larger framework of power, such as a colonial governance or a national primary education agenda. Consequently, to borrow from Arjun Appadurai, we as modern individuals encounter a "plurality of imagined worlds."[32] As shown here, to conceive of this plurality as indexed by American adults and children at the end of the nineteenth century, one needs to consider that different groups of them referred to divergent real or imagined worlds on both the individual and the collective levels. Consequently, in acts of imagination, policy making, instructing, writing, reading, and playing, Americans—adult and child—did not allude to a *single* non-American world but in fact a plurality of worlds. Ultimately, as I argued before, mapping and placing the United States in the world at large constituted a plethora of individual and national, imaginative, and scientific projects, each of which conceived of the nation as occupying a slightly different place on the map of a slightly differently imagined world. In light of this, and this is a central argument in *Citizens and Rulers of the World*, geographic knowledge is provisional, puzzling, and, most importantly, plural.[33]

To return to Tara Zahra's observations on modern nations, I believe that terms such as "national disinterest" and "national ambivalence" register the potential disappointment and pejorative connotations that politicians have historically ascribed to the lackluster patriotism of enfranchised adult citizens.[34] In light of the discussions made in the past few pages, a different term, then, needs to be devised in reference to children's responses to world geography; it is in effect more apt, I conclude, to view children's relationships to adult projects to ensure their future as stewards of a global empire as aesthetic, spirited, cartography-in-progress attempts to offset the confusion generated by the immensity of the data about the world(s) that they were fed through primers, games, the periodical press, and conversations with other children and adults. After all, while American politicians and entrepreneurs promised and planned to render the U.S. empire a spatially mappable entity, children responded to those promises and plans with individual, impetuous acts of cognitive mapping.

To conclude, if unaware of the basics of cartography and the blatantly reductionist spatial arrangement of relationships between towns, countries, and continents; if uninterested in the complex colonial histories of the place names they asked about or looked for in order to compose and solve geographical puzzles; if leading localized, domestic lives in the American countryside; if too poor or too ill or too remotely located to travel abroad, children still encountered world geography in contexts beyond the classroom—even outside the domain deemed proper to geography as an area of study. American children's immediate home environment, leisure activities, and educational entertainment—in flux at the pivotal intersection of religion, gender, class, ability, and ethnicity, among other factors—provided them with a myriad of modes of encountering and making sense of the world. Many turn-of-the-century American children had, in fact, made it their competitive task to think of their home as one among many in the world, to name and list various parts of the world in geographical puzzles, and to address the tensions between their "national ambivalence" and the expansionist intents of the adults around them by exercising mastery over world geography during countryside picnics or while reciting their alphabet.[35]

Notes

Introduction

1. Written in the same spirit as Jules Verne's popular *Five Weeks in a Balloon*, published in 1863, *Tom Sawyer Abroad* (1894) took Twain about a month to write. The first of an unfinished travel book series for boys, the book was first published in six installments from November 1893 to April 1894 in the juvenile periodical *St. Nicholas* before it appeared as a book based on the manuscript edited by *St. Nicholas*'s editor, Mary Mapes Dodge. It was republished together with *Tom Sawyer, Detective* (first serialized in the *Harper's New Monthly Magazine* from August to September 1896) as *Tom Sawyer Abroad; Tom Sawyer Detective; and Other Stories*. We know from Twain's autobiography that he had experienced a balloon ride together with the then-governor of Wisconsin and a few other friends in June 1879. For a detailed analysis of Twain's fiction in relation to his own life events, see the editors' introduction to Twain, *The Adventures of Tom Sawyer*, edited by John C. Gerber, Paul Baender, and Terry Firkins.

2. In the story, the boys make it all the way to the Sahara Desert and Egypt, where they visit the pyramids and the Sphinx. Twain, *Tom Sawyer Abroad*.

3. For critical reflections on Mark Twain's *Tom Sawyer Abroad* in geographical terms, see Haggett, *Geography: A Global Synthesis*. For a fine-grained study of Twain's fascination with geography in both fiction and travel writing, see Marx, "Pilot and the Passenger"; Alvarez, *Mark Twain's Geographical Imagination*.

4. Tom's and Huck's naive trust that maps do not tell lies counters the long-established fact that, in their attempt not only to reduce, connect, or convert dimensions but also to frame, fabricate, or forge unfounded spatial claims, maps lie all the time and in a host of trivial and consequential contexts. For a fascinating account of maps as a medium of deception, see Monmonier, *How to Lie with Maps*.

5. For a compelling survey of the topic in post-Enlightenment map markets and print cultures of Europe, see Verdier and Besse, "Color and Cartography," 294–302. For detailed observations on the advances in U.S. cartography, including the adoption of chromolithography during the course of the nineteenth century, see chapters 3 and 4 in Brückner, *Social Life of Maps in America*. See also Bosse, "To Give a Strong and Pleasing Effect," 32–37.

6. Similar notions of mapping and scalar tension appear in Jorge Luis Borges's 1946 short story "Del rigor en la ciencia," set in 1658. In the story, Borges recounts the mapping of an imaginary empire on a one-to-one scale. The scheme, the story reveals, results initially in self-aggrandizement and exactitude, and later in apprehension and dismissal of the map because it is too cumbersome to study. Borges and Hurley, *Collected Fictions*.

7. Brückner, "Lithographed Map in Philadelphia," 160–61.

8. Jean Baudrillard, "Simulacra and Simulations," 166.

9. Thongchai Winichakul's views on the colonization of Siam that appeared in his doctoral thesis in 1988 are expanded on by Benedict Anderson in his examination of the material sites

of colonialism. For a full account of Anderson's adoption of Thongchai Winichakul's ideas, see "Census, Map, Museum." See also Thongchai Winichakul, *Siam Mapped*, 130.

10. By the time *Tom Sawyer Abroad* came out, Twain had already established himself as far more than a literary figure. He was no outsider to debates on politics, geography, and imperialism. Twain's fascination with these topics is apparent in a wide range of his works, especially in lectures he gave on his tours of Hawai'i, the United States, and England, and later in his vociferous anti-imperial essays, such as "To the Person Sitting in Darkness" (1901), "The War Prayer" (1905), and "King Leopold's Soliloquy" (1905). Twain's long-standing interest in travel and human movement through space catered to the general tendency among Americans to read him as an authority who teased their curiosity and fed their fascination with the reciprocal relationship between geography and politics as an anchor for the changing place of the nation within, and in relation to, the world. For an analysis of Twain's career as a travel writer and the influence his travel writing had on his less well-traveled nineteenth-century American audiences, see the introduction to Melton, *Mark Twain, Travel Books, and Tourism*. For a thorough examination of the notions of race and imperialism and of the nativist sentiments at play in Twain's works, see Hsu, *Sitting in Darkness*.

11. The phrase "citizens and rulers of the world" appears in an essay that Albert Perry Brigham wrote in 1897 about secondary school geographic pedagogy and the opening of U.S. geography to the world. For further details, see chapter 1.

12. An exhaustive examination of late nineteenth-century adventure literature is beyond the scope of the present book, which instead focuses on nonliterary archival material. However, geographical fiction, such as *Tom Sawyer Abroad*, provides a fruitful lens to the study of the ways geographical education and knowledge were indexed in cultural products beyond the narrow realm of formal pedagogy around the turn of the twentieth century.

13. For compelling examinations of the many domestic and global changes that the decade stands for in U.S. history, what Brands has termed "the reckless decade," see, among others, Smith and Dawson, *American 1890s*; Brands, *Reckless Decade*; Hamilton, *America's New Empire*; Healy, *US Expansionism*.

14. For a considered discussion of the unique path the United States took toward nationhood, see King, *Liberty of Strangers*. King proposes that American nationhood began as an internal struggle over Americanization, at the heart of which had always lain a "frontier crisis."

15. Jehlen, *American Incarnation*, 9.

16. Friedrich Ratzel wrote extensively on state politics in conversation with geography and other spatial considerations such as weather, natural resources, and evolution. The term "geopolitics," however, predates his writings. Geopolitics had in fact been first adopted by Rudolf Kjellen in reference to Swedish-Norwegian border conflicts in 1899, after which it was exported to other national frameworks for thinking about state politics. The notion was extensively complicated by European as well as American geologists and geographers such as American geographer Ellen Churchill Semple, American geologist William Morris Davis, and American geographer and educator Richard Elwood Dodge. Ratzel himself was influenced in his writings by the American naval geo-strategist Alfred Thayer Mahan, whose insights Ratzel adopted as justification for imperial expansionism through sea power. For an exhaustive history of the term, including its definitions and uses over time, see Stein-

metz, "Geopolitics." For an examination of the geopolitically significant term *Lebensraum*, which Ratzel popularized and which was later adopted by the National Socialist Party in the prewar years, see W. Smith, "Friedrich Ratzel and the Origins of Lebensraum." The roots and genealogy of geopolitical thought in the United States, including the influence Ratzel had on it, are examined at length in N. Smith, *American Empire*, especially chapter 10 on geopolitics.

17. Toal, *Critical Geopolitics*, 2.

18. By the dawn of the new century, anthropocentric approaches to geography were so common that Ratzel subtitled his two-volume *Anthropogeographie* with "*oder Grundzuge der Anwendung der Erdkunde auf die Geschichte.*" The first volume of *Anthropogeographie* was published in 1882. The second volume appeared in 1891 and resulted in heated debates and engaging reviews on both sides of the Atlantic. Ratzel's understanding of geography had a strong influence on many of his contemporaries in the United States, such as Ellen Churchill Semple, and J. Russell Smith, who studied under Ratzel in 1901. Beyond the discipline of geography, economists and political enthusiasts, too, were interested in Ratzel's ideas. His works were read and reviewed in the United States by academic economists, anthropologists, sociologists, and political economists such as William Z. Ripley (1867–1941), a professor of political economics and anthropology at the Massachusetts Institute of Technology and later at Harvard University (1894); Charles Richmond Henderson (1848–1915), a professor of sociology at the University of Chicago (1903); Jesse Walter Fewkes (1850–1930), an American archeologist and anthropologist (1897); and Ratzel's American disciple Ellen Churchill Semple (1863–1932), herself a proponent of human geography and environmental determinism, and later a professor at the University of Chicago (1894 and 1900). While Semple adopted Ratzel's views to a great degree, Ratzel was so influenced by contemporary American thought on geography that, paying homage to Alfred Thayer Mahan's *Influence of Sea Power upon History 1660–1783*, he penned *The Sea as a Source of the Greatness of a People* in 1900.

Susan Schulten closely examines the exchanges, both direct and indirect, that took place between Ratzel, Semple, William Morris Davis, Alfred Thayer Mahan, and Theodore Roosevelt. Schulten, *Geographical Imagination in America*, 79–82. See also Holt-Jensen, *Geography: History and Concepts*, 63–65.

19. On the reciprocities of education and citizenship in United States between 1882 and 1924, see the rigorously focused study *Education for Empire* by Clif Stratton.

20. Mackenzie, "Report of the Committee of Ten," 146.

21. On the evolution of the notion of childhood the way it is commonly understood today in the West, see, among others, Fass, *Routledge History of Childhood*; Cunningham, *Children and Childhood in Western Society*; Heywood, *History of Childhood*. See also the foundational, much-debated classic volume *L'Enfant et la vie familiale sous l'Ancien Régime* by Philippe Ariès, translated by Robert Baldick and published in English as *Centuries of Childhood*. Ariès's work has been read and referenced widely by professional historians and childhood studies scholars. For decades now, his views on childhood have been met with waves of enthusiastic adoption, well-grounded dismissal, and continued cautious referencing. Ariès argues that childhood, as a category separate from adulthood, was only "invented" after the Middle Ages. This and other arguments in the volume—including his use of high art as evidence and his placement of the origins of the Western notion of childhood to

early modern times, as if children did not matter to adults before then—have haunted the study of childhood to this day. As Anastasia Ulanowicz suggests, some controversies surrounding Ariès's work have resulted from the mistranslation of a number of key terms in his book; Ulanowicz, "Philippe Ariès." On the other hand, a wide number of prejudiced or poorly informed arguments that he makes, such as passing claims in the chapters "From Immodesty to Innocence" (100–127) and "Medieval Scholars: Young and Old" (137–54) about sexuality and child education in Muslim societies, have remained by and large uncriticized.

22. Hawes and Hiner, "Reflections on the History of Children and Childhood," 24.

23. While childhood is both a biosocial developmental stage of human life and a sociocultural construct imagined and invoked by adults, especially in times of sociopolitical crisis or national urgency, I employ childhood in the present volume mainly as the latter (which was the predominant view of childhood at the time under study here): a toy, an artifact, and a sheltered cosmos that adults created and gave to children in the hope that they would obediently play with, learn from, and outgrow it.

24. A deeper examination of the history of historical childhood studies as a discipline resides beyond the scope of the present volume. For a thorough and accessible survey of the discipline's origins and transformations over the past half century, see, among others, Moruzi, Musgrove, and Leahy, "Hearing Children's Voices," 1–25.

25. *Making of a Great Magazine*, 10.

26. For a deeper understanding of the various incarnations and the history of the phrase "young America"—as a catchphrase, a rhetorical tool to downplay the significance of the study of U.S. history, an 1830s and 1840s New York–based literary movement to promote the production of highbrow American literature, and a political faction of the Democratic Party most popular during the 1840s and 1850s—see Eyal, *Young America Movement and the Transformation of the Democratic Party*, especially the introduction.

27. G. S. Hall, *Adolescence*, xvi.

28. For contemporary, multidisciplinary studies that consider children and childhood in spatial terms, see, among others, Hackett, Procter, and Seymour, *Children's Spatialities*; Alasuutari, Mustola, and Rutanen, *Exploring Materiality in Childhood*.

29. The phrase "the family of nations" is a variation of a rather common nineteenth-century expression adopted in discussions about the interactions among sovereign states on an international scale. McKinley's rhetorical use of the phrase in the middle of the nation's colonial involvement with the Philippines was meant to mark the rise of the United States as an emerging overseas empire to an equal standing with those of the older imperial members of this family. Noteworthy is the fact that while pre-1870 geography primers emphasized and celebrated the young nation's entry into the "family of nations" as justification for the westward expansion of the nation, this rhetoric was reinvoked at the turn of the twentieth century to endorse the nation's global enterprises. "William McKinley: Second Inaugural Address," 257.

30. Nodelman, "Orientalism, Colonialism, and Children's Literature," 29–35.

31. See, for instance, Grieshaber, *Rethinking Parent and Child Conflict*; Ashcroft, *On Post-Colonial Futures*; Rollo, "Feral Children." There is an internal conflict in the way the colonized person is read as a child—at once endearing (and convenient) in their dependence and alienating (and repulsive) in their inferiority: "The child is primitive, pre-literate, edu-

cable, formed and forming in the image of the parent. There are no colonies which are primitive without being childlike in their amenability to instruction; there are no colonies which are sons and daughters of empire without being marginal, negated and debased to some extent. This is because the very existence of empire itself rests upon the security of its binary logic of centre and margin" (Ashcroft, 47–48).

32. Sánchez-Eppler, *Dependent States*, xvi.

33. Crain, *Story of A*; Crain, *Reading Children*. Read the latter work together with Margaret Mackey's autobiography, a volume in which she revisits her own childhood and the ways her life shaped and informed her reading self in mid-twentieth-century Newfoundland. Mackey, *One Child Reading*.

34. Crain, *Story of A*, 4.

35. For thorough historical observations on the evolution of geography and cartography in the United States, see Brückner, *Social Life of Maps*; N. Smith, *American Empire*; Schulten, *Geographical Imagination in America*; Schulten, *Mapping the Nation*. Schulten's *Geographical Imagination in America*, for one, sheds much needed light on the evolution of school geography along the axis of geopolitics from 1880 to 1950—a topic on which I expand in chapter 1. Schulten's chapter "School Geography, the 'Mother of All Sciences,' 1880–1914" discusses the modernization of geography as a discipline, its turn to the human element and "Social Darwinism," and how this new science was taught through revised school primers before World War I. The following chapter, "School Geography in the Age of Internationalism, 1914–1950," further examines the changes in the nature of school geography between the two world wars.

36. Though lying outside the scope of discussions in the present book, a notable volume that examines American childhood as an ambiguously entertaining and didactic, adult-bred artifact is Onion, *Innocent Experiments*. Keeping an eye throughout the volume on how childhood was perceived during the twentieth century as invariably cute, innocent, and curious, Onion investigates the gendered, domesticated, and mostly extracurricular nature of the so-called popular science as enacted by American children.

37. Norcia, "Playing Empire." See also her more recent work, *Gaming Empire in Children's British Board Games*. By the same token, Ann McGrath's article "Playing Colonial" focuses on the twentieth-century heritage of the British Empire as consumed by Australian and North American children. In her comparison of the variations of "Cowboys and Indians" games in Australia and North America, she traces "a cultural pairing of empires of the imagination where history became performance and where a global modern identity was installed, historicized, and contested by both children and adults." McGrath, "Playing Colonial," 2.

38. Wong, "Around the World and across the Board; Norcia, "Puzzling Empire: Early Puzzles and Dissected Maps as Imperial Heuristics."

39. In *Children, Childhood and Cultural Heritage*, Kate Darian-Smith and Carla Pascoe point out the "universal presence" and the resultant invisibility of children in the world of adults, a logical consequence of which was the scarcity of research into children's historical presence and heritage until about two decades ago. Darian-Smith and Pascoe, *Children, Childhood and Cultural Heritage*, 2.

40. Engaging with The question of historical presence in the present volume, I echo Kristine Alexander's and Mona Gleason's forceful invitations to historians of childhood to

reconsider what work "child agency" does beyond reproducing the binarism long at work between children and adults. See Alexander, "Agency and Emotion Work"; Gleason, "Avoiding the Agency Trap." In "Agency and Emotion Work," Alexander rightly criticizes the term's intellectual roots in European Enlightenment as it is often coupled with such Enlightenment notions as progress and individual freedom (121). She further calls for asking more meticulous questions when we encounter children of the past (or traces of their presence in records) than merely repeating questions about whether and how they exercised agency: "What nuances and specificities," Alexander asks, "are lost when studies of young people in vastly different times and places all base their analyses on the uniform claim that their subjects had some kind of agency?" (123). See also Miller, "Assent as Agency"; Vallgårda, Alexander, and Olsen, "Against Agency."

41. Sánchez-Eppler, *Dependent States*, xxiv.

42. More recent historical accounts—such as Weikle-Mills, *Imaginary Citizens*; Cohoon, *Serialized Citizenships*; Smith and Duane, *Who Writes for Black Children?*; and Field, *Playing with the Book*—join Patricia Crain's work as remarkable proof of success against these older trends.

43. Stoler, "Tense and Tender Ties"; Hoose, *We Were There, Too!*; Mintz, *Huck's Raft*. Mintz's *Huck's Raft*, for instance, bases its study of African American, white, and Native American childhoods during and after colonial times almost entirely on retrospective or legal sources that either are based on childhood memories or point to lived experiences of child labor, child-rearing, and parenting. Pursuing a closely related methodological model, *Citizens and Rulers of the World* assembles strands of thought from across childhood studies in order to prioritize children's social and meaning-making roles in the American age of empire.

44. Bowersox, *Raising Germans in the Age of Empire*. Discussing the role the Boy Scouts of America played in molding imperial masculinities in the United States during the first half of the twentieth century, Mischa Honeck pays similarly keen attention to American children as potential citizens and rulers of the world at the dawn of the twentieth century. Honeck, *Our Frontier Is the World*. From a strictly gendered perspective, Benjamin René Jordan argues that the BSA was, from the outset, a project of redefining American manhood for an increasingly diverse demographic target group across urban America. Jordan, *Modern Manhood and the Boy Scouts of America*.

45. Careful multifaceted discussions of how women entered historiography in the West—the challenges this posed to the established masculine canon of historians and their topics of choice (what Epple and Schaser refer to as "a real troublemaker for historiography as a whole"), the question of women's archives, and the ways gender-marked history changed the contours of history at large—appear in Scott, *Gender and the Politics of History*; Butler and Weed, *Question of Gender*; and the introduction to Epple and Schaser, *Gendering Historiography*.

46. For a comprehensive analysis of the birth of the United States as a nation, its infancy (the birth of liberal ideals), and how those ideals shaped early American identity in terms of geography and liberalism, both symbolically and ideologically, see Jehlen, *American Incarnation*.

47. While the United States covered 888,811 square miles in 1800, its government pushed for expansion and the country grew throughout the century as a result of conquest, pur-

chase, war, and the aggressive settlement of massive stretches of land. Consequently, by 1900, the United States reached across four time zones, touched two oceans, and covered a land area of 3,022,387 square miles.

48. Hsu, *Geography and the Production of Space*, 1.

49. Colin Woodard offers an insightful foray into colonial times to suggest that despite white America's claims to national unity, the United States was, and indeed still is, a nation of disparate rival communities, each with its own set of ideas, values, and practices. Woodard, *American Nations*. Furthermore, Thomas Bender's seminal work *Rethinking American History in a Global Age* examines nationhood and nationalism in the United States in a transnational context. In the introduction to the book, Bender discusses how white, mainly middle-class, America tried to disentangle and push aside other equally significant and valid narratives throughout the course of the nineteenth century, laboring tirelessly, even at times aggressively, to forge a unified, albeit primarily white, national narrative—the e pluribus unum—in order to get Americans of diverse backgrounds to subscribe to it.

50. It should be noted that the U.S. imperial scenario was different from that of European empires in that migrants from various parts of the world arrived at its shores almost daily, making white Americans' aggressive colonial claims to national purity and geographical sovereignty almost impossible to sustain. For groundbreaking discussions on territorial expansion and overcoming geographical distances as key imperial movements in the age of empire, see Said, *Culture and Imperialism*. For a finespun discussion of the territorial expanse of the U.S. empire as an archipelagic sphere of influence, see Roberts and Stephens, "Archipelagic American Studies."

51. Roberts and Stephens, "Archipelagic American Studies."

52. Roberts and Stephens, "Archipelagic American Studies," 1.

53. Delivered on March 4, 1897, President McKinley's first inaugural speech focused on the more urgent domestic issues of the time. The speech echoed Washington's isolationist approach and Monroe's non-entanglement doctrine to the full. As previously discussed, this was in contrast to his second inaugural speech, delivered four years later. "William McKinley: Second Inaugural Address." At the same time, McKinley's position in this speech stood in contrast to the aggressive expansionist views promoted by Republican politicians such as Henry Cabot Lodge and McKinley's assistant secretary of war and successor as the twenty-sixth U.S. president, Theodore Roosevelt. On Roosevelt's views, see, for instance, "Expansion and Peace," first published in December 1899 in the *Independent*, where Roosevelt aggressively coupled peace with war and argued that cowardice in an unjust peace is a greater national betrayal than valor in a just war. In defending the pro-expansionist route paved by the Spanish-American War, he further observed that "fundamentally the cause of expansion is the cause of peace." Roosevelt, *Strenuous Life*, 34. For a fine roundup of Roosevelt's expansionist views, see Watts, *Rough Rider in the White House*.

54. "William McKinley," 257.

55. "William McKinley," 256.

56. "William McKinley," 257.

57. "William McKinley," 257.

58. Withers, *Placing the Enlightenment*, 10. Examining geographies of the Enlightenment, Withers applies the term "geographically privileged persons" to two groups of people who,

during the course of the eighteenth century, produced geographic knowledge for public consumption: first, individuals—usually young, able-bodied white males—who traveled extensively and offered firsthand accounts of what they saw, and second, those individuals—older adults, mainly white, usually male but occasionally female, as was the case with late nineteenth-century professional geographers—whose privileged social backgrounds and access, both direct and indirect, to geographic knowledge made them trustworthy sources of such knowledge in the public eye. I adopt the term, stretching it in reference to late nineteenth-century adults who produced geographic knowledge both for children and for other adults. For a fascinating account of nineteenth-century British geography primers written by female authors, see Norcia, *X Marks the Spot*.

59. However, this does not mean that turn-of-the-century Americans were more *accurately* familiar with or keener on world geography than they had been earlier or were going to be later. In fact, Americans were never more map literate and geography savvy than during World War II. For illuminating discussions on geographic literacy during the period 1880–1950, see Schulten, *Geographical Imagination in America*.

60. Quoted in N. Smith, *American Empire*, 7. For a comprehensive account of *National Geographic*'s traction as "a cultural icon" as well as "a generator of icons" since its inception, as a timely venue for popularizing geographical imagination in 1888, and as a motor of U.S. foreign policy since the outbreak of the Spanish-Cuban-Philippine-American War of 1898, see, among others, Hawkins, *American Iconographic*; Lutz and Collins, *Reading National Geographic*. On the anthropological and orientalist undertones of the magazine, see Steet, *Veils and Daggers*; Little, *American Orientalism*; Rothenberg, *Presenting America's World*.

61. Ford, *Issues of War and Peace*, 165.

62. Library of Congress, "History of the Library of Congress."

63. Brückner, *Early American Cartographies*, 6.

64. For a fine-grained analysis of how cartography as a European colonial import made its way to Americans' everyday and academic life and multiplied in its uses within the United States during the nineteenth century, see Schulten, *Mapping the Nation*; Schulten, *Geographical Imagination in America*. For a comprehensive survey of the role geography played throughout the first 150 years of the nation, see Brückner, *Geographic Revolution in Early America*. Moreover, Phillips's *Mapping Men and Empire* offers a fascinating account of mapping adventures and the gendered aspects of geography. Finally, Brückner and Hsu, *American Literary Geographies*, provides a rich and diverse series of essays on geography and American literary history from 1500 to the dawn of the twentieth century.

65. Twain, *Innocents Abroad*, chap. 1.

66. The term "picnic" seems first to have appeared in the 1692 edition of *Les origines de la langue françoise de Ménage*, suggesting that it was a common term (and practice) among the French at least during the second half of the seventeenth century. It was later adopted in English with roughly similar connotations. Ménage, *Les origines de la langue françoise de Ménage*. On the other hand, Thomas Cole's 1846 picnic painting is believed to have been the first artwork to bring the idea of picnicking into Americans' popular imagination. Cole painted *A Pic-Nic* at the height of his fascination with landscape. Cole is most famous for his series *Course of Empire* (painted between 1833 and 1836), which has been invoked and analyzed in historical studies of the empire, such as Hsu's *Geography and the Production of Space*. See Cole, *A Pic-Nic Party*.

67. According to Sánchez-Eppler, in the last quarter of the nineteenth century, the United States had sent the most Protestant missionaries of any other Western Christian countries to the non-American world. Sánchez-Eppler, *Dependent States*, 187.

68. Jameson, "Cognitive Mapping," 350.

69. For a classic interpretation of the economic roots of the U.S. rise to "world power," see LaFeber, *New Empire*. There, LaFeber explains that the industrial revolution in the second half of the nineteenth century was the key cause of a number of fundamental changes in U.S. society that ultimately led to social unrest, economic depression, and a political urge to look for both raw material and overseas markets for the surplus product of American factories. LaFeber's book is one among many attempts by Cold War historians and political scientists to explain the rise of the United States to a global empire at the turn of the twentieth century in a more or less benevolent light. For a comprehensive debate on these factors, see the volume edited by McCoy and Scarano, *Colonial Crucible*, especially the articles in part 1, "Exploring Imperial Transitions."

70. Harvey, *Condition of Postmodernity*, 302; Jameson, *Postmodernism*, 44.

71. The European origins of the Mercator projection (engraved by Gerard Mercator and published by Abraham Ortelius), its production during the heyday of the European age of exploration, and its massive popularity on both sides of the Atlantic (first produced in 1569, published as a stand-alone map in 1587, and included in the Mercator Atlas in 1595) had transformed it into a common source of geographical imagination and knowledge production based on European—mainly imperial—conceptions of the world which are still at work to this day. For a careful discussion on the Mercator projection's continued popularity, the novel ways it was put to use, and the criticism leveled against it in the nineteenth century, see Monmonier, *Rhumb Lines and Map Wars*. For criticisms leveled against the Mercator projection at the turn of the twentieth century and the rise of competing projections by American cartographers, see Schulten, *Geographical Imagination in America*.

72. Jameson, *Postmodernism*, 409.

73. Jacob, *Sovereign Map*, 360.

74. For Lynch's original thoughts on the concept, see Lynch, *Image of the City*. For Jameson's reading of Lynch's mental mapping, see Jameson, *Postmodernism*, 51–54.

75. Compare Lynch's understanding of cognitive maps with Peter Gould's notion of "mental maps," proposed in the 1970s and 1980s. In a recent return to the study of mental mapping, Gould reemphasizes the centrality of the practice in human geography, asserting that while we know little about how mental maps and spatial images form and change in people's minds, mental maps are key to the study of the *human* element of the field. See Gould, "On Mental Maps," 182. For his original work on mental maps, see Gould, *Mental Maps*.

76. For this critique of Jameson's notion of cognitive mapping, see Colin MacCabe, preface to *The Geopolitical Aesthetic*; Hale, "Cognitive Mapping."

77. Other equally urgent questions include: How do factors such as colonization, forced migration, and displacement affect individual cognitive maps and call for other maps and other modes of being in the world? And, as a result of such spatial exclusions and trauma, what unforeseen spatial identities form?

78. Harvey, *Condition of Postmodernity*, 220.

79. Harvey's attempt at theorizing the ways he reads capitalism and its power to determine, alter, but also destroy the world can be followed in a number of books he published

in the 1980s and 1990s. See, among others, Harvey, *The Limits to Capital*; Harvey, *Condition of Postmodernity*. See also Harvey, "Between Space and Time."

80. Harvey, *Condition of Postmodernity*, 220–21.

81. As Harvey himself contends, to consider his grid as a final list of spatial practices "would be to accept the idea that there is some universal spatial language independent of social practices." Harvey, *Condition of Postmodernity*, 222–23.

82. Ginzburg, "Clues."

83. Epple, "Globale Mikrogeschichte."

84. Ginsburg, "Clues," 88.

85. Ginsburg, "Clues," 92–97.

86. For a broader understanding of microhistory as methodology, see Peltonen, "What Is Micro in Microhistory?"

87. For an understanding of relationality in global history writing, see Epple, "Globale Mikrogeschichte."

88. Charles W. J. Withers tackles the question of modern geography and geographic-mindedness of the Enlightenment in the long eighteenth century, dating the birth of modern geography to the Enlightenment. For his take on the place of geography in the Enlightenment and the role the Enlightenment played in changing the nature and focus of geography, see Withers, *Placing the Enlightenment*. For foundational studies that address this topic and on whom Withers builds his arguments, see, among others, Bowen, *Empiricism and Geographical Thought*; Livingston, *Geographical Tradition*.

89. G. S. Hall, *Educational Problems*, 555–56.

90. Gordon Kelly lists over 280 children's magazines that were launched in the United States between 1789 and 1899. Kelly, *Children's Periodicals of the United States*, 553–59.

91. Mott, *History of American Magazines*, 273.

92. Kelly, *Children's Periodicals of the United States*, xxv.

93. Pointing to Boston's late-century failure to ably compete with New York as the nation's cultural capital, John Tebbel and Ellen Zuckerman mention the long-lived juvenile periodical *Youth's Companion* (with the remarkable circulation number of about 385,000 by 1885) as the only magazine that the nation's former cultural capital continued to publish. Though it was undoubtedly a well-liked children's magazine that shaped the taste of generations of its readers both before and after the Civil War, Tebbel and Zuckerman describe the *Youth's Companion* toward the end of its life as "scarcely cultural." Tebbel and Zuckerman, *Magazine in America*, 58.

94. Kelly and J. D. Stahl, among others, name *St. Nicholas* as the best American juvenile periodical, especially during the 1880s and 1890s. Kelly, *Children's Periodicals of the United States*, 377–78; Stahl, "Children's Literature," 221. As two of the most popular children's magazines of the period, *St. Nicholas* and *Harper's Young People* have been extensively studied by both historians of childhood and literary critics. See, for instance, Redcay, "'Live to Learn and Learn to Live'"; Gannon, "'Best Magazine for Children of All Ages'"; Gannon, Rahn, and Thompson, *St. Nicholas and Mary Mapes Dodge*; Cane and Alves, *Only Efficient Instrument*; Clark, *Kiddie Lit*; McKenzie, "A 'Revolutionary' War?"

95. *Scribner's Monthly* heralded the arrival of *St. Nicholas* in 1873 in its column Topics of the Time: "Whether we shall lead the little child, or the little child shall lead us, remains to be seen; but it will be pleasant to have him at our side, to watch his growth and develop-

ment, and to minister, as we may, to his prosperity.... What more can be said of it, except to assure fathers and mothers and children everywhere that they will want it, and must have it. Wherever 'Scribner' goes, 'St. Nicholas' ought to go. They will be harmonious companions in the family, and the helpers of each other in the work of instruction, culture and entertainment." "St. Nicholas," *Scribner's Monthly*, 115.

Chapter One

1. Webster, *On the Education of Youth in America*, 1.

2. Although it is left unacknowledged in Morse's volume, it is highly probable that at least some of the "mistakes" that Morse referred to had been brought to life by female authors of British geography primers—authors whose works were met with hostility among male geographers of the time. For a fascinating account of the long nineteenth-century geography textbooks of the British Empire that were researched and written by professional female authors, see Norcia, *X Marks the Spot*.

3. J. Morse, *American Geography*, 1. Given the sensitive nature of the post-Revolutionary era in which Morse's American geography was published, note the terminology which Morse deployed in describing the United States: "an independent nation" and an "empire." Further note that his *American Geography* did in fact include no map of the United States as part of a larger world.

4. Winthrop, "Model of Christian Charity," 191–93; Wood, *Empire of Liberty*, 3. For a thoroughgoing survey of the celebratory mood in the early republic in the middle of tensions between American political leaders' claims to parochialism and their aspirations to cosmopolitanism, see Wood, *Empire of Liberty*, chap. 1.

5. Wright, *Geography in the Making*, provides a compelling historical discussion of the American Geographical Society during the first century of its existence. Preston E. James and Clarence Fielden Jones offer an insightful analysis of the state of geography in the United States and of the role the American Geographical Society and the United States Geological Survey played in its evolution in *American Geography*.

6. Schulten, *Geographical Imagination in America*, 101.

7. The pre-national roots of American exceptionalism are well documented. In one of the best remembered religious sermons of the past four centuries, John Winthrop's "Model of Christian Charity" was composed primarily to justify the expansion of British rule in the "New World." While hoping to dutifully prepare the colonists for what they were going to encounter in the colonies, Winthrop's promise of "a city upon a hill" actually set the tone for an exceptionalist rhetoric that would repeat itself over the centuries. Winthrop, "Model of Christian Charity," 191–93. Over time, Winthrop's vision gave rise to a variety of discourses of "exceptionalism," reinvoked, for instance, by John L. O'Sullivan, who, in 1839, wrote about the roots of "our disconnected position as regards any other nation." Best known as the journalist who came up with the phrase "manifest destiny" in 1845, O'Sullivan captured existing public sentiments regarding isolationism at the heart of U.S. foreign policy as a matter not of political choice but of national character. O'Sullivan, "Great Nation of Futurity." Donald Trump's "Make America Great Again" is one of the latest, most simplified iterations of that same mentality.

8. D'haen, Giles, Kadir, and Parkinson Zamora, *How Far Is America from Here?*, 14.

9. Among the numerous calls for reformed approaches to studying the United States outside of a national frame of reference, one major strand has been developed by scholars at the Center for InterAmerican Studies (CIAS) at Bielefeld University. The interdisciplinary project "The Americas as Space of Entanglements" has provided a historically informed and spatially substantiated theoretical framework within which Americanists have conducted research in order to examine the United States, first as one among many political entities in the Americas, and second as a spatial entity on the map of a complexly interconnected and densely imagined world of dynamic entanglements. For details about the project, see Center for InterAmerican Studies, "Entangled Americas."

10. Bender, *Nation among Nations*, 4.

11. Epple, "The Global, the Transnational and the Subaltern," 155. Note here that Bender's observation that nations cannot be their own historical context serves as the temporal equivalent of Epple's predominantly spatially informed argument.

12. Bender, *Nation among Nations*, 4.

13. Pratt, "Science, Planetary Consciousness, Interiors," 15.

14. For an earlier parochial incarnation of the notion of "home geography" early in the nineteenth century, see Woodbridge and Willard, *System of Universal Geography*. For the Pestalozzian roots of their views toward the study of geography, see Tröhler, *Pestalozzi and the Educationalization of the World*.

15. As I noted in the introduction, geography's modernization at the turn of the twentieth century had numerous contours, including the development of a scientific vocabulary (which can be traced in the works of prominent geographers such as William Morris Davis) and the introduction of theories such as Lamarckian evolution into the discipline's language (which had formerly consisted mainly of observation). For a detailed account of these changes, see Schulten, *Geographical Imagination in America*, especially chaps. 4–6. For further fascinating discussions of other, equally significant aspects of this modernization process, see Dunbar, *Geography*; N. Smith, *American Empire*; Brückner and Hsu, *American Literary Geographies*; Rankin, *After the Map*.

16. Though first published in 1795 in Hartford, Connecticut, the edition I consulted in writing this chapter was published in 1808 in New York. Dwight, *Short but Comprehensive System*.

17. Dwight, *Short but Comprehensive System*, 108.

18. Dwight, *Short but Comprehensive System*, 139.

19. Dwight, *Short but Comprehensive System*, 98.

20. Brückner, "Lessons in Geography," 315. For a more general examination of school books as performative of the national and the patriotic in the early decades since the Declaration of Independence, see G. A. Adams, "'Pictures of the Vicious.'"

21. Brückner, "Lessons in Geography," 315.

22. Adams, "Pictures of the Vicious,'" 152. On the role literacy campaigns played as nationalist and patriotic projects in the United States, see Crain, *Story of A*. On the crucial ways geographic writing helped foster the prerequisite rhetoric and imagery adopted to advance these projects, see Brückner, *Geographic Revolution in Early America*.

23. O'Sullivan, "Annexation," 5.

24. In *Old-Time Schools and School-Books*, Clifton Johnson points to the accompanying terrestrial globes and atlases that he believes are not to be found in archives due to their low

quality and overuse by young owners. According to him, the atlases "were flimsily made, with paper covers, and the wear and tear of daily use made an end of them." Clifton Johnson, *Old-Time Schools and School-Books*, 337. Another example of table globes that did not survive the wear and tear of frequent use in the schoolroom is what globe makers referred to as "slated globes." For an examination of these blank black globes in the context of the pedagogies of empire, see my "What on Earth!," where I read slated globes as "practice fields of world geography." Mayar, "What on Earth!," 16. A locus classicus for the study of the relationship between memorization and geographic literacy is Brückner, *Social Life of Maps*, especially chaps. 7, 8.

25. Brückner, *Social Life of Maps*, 337–38.

26. For a thorough examination of the pedagogic and patriotic uses of atlases and outline maps in the early republic, see Brückner, *Social Life of Maps*, esp. chap. 7.

27. Carpenter, *History of American Schoolbooks*, 258.

28. D. Adams, *Modern Geography*; Clute, *The School Geography*. Next to these volumes, Harriet Beecher Stowe's *First Geography for Children* is a striking "activist" textbook that directly engaged with slavery, questioned racial hierarchies at the heart of U.S. geography teaching, and taught young American children about abolitionism.

29. Goodrich, *Peter Parley's Method of Telling about Geography*. See also Goodrich's other Peter Parley geography books (listed in the bibliography).

30. Prominent American educators such as Henry Eldridge Bourne believed that history and geography should be taught together, as they were inseparable—particularly when it came to accounts of Europe in the age of exploration. Eldridge Bourne, *Teaching of History and Civics*, 112.

31. Goodrich, *Peter Parley's Method of Telling about Geography*, v.

32. See also *Heartless Immensity*, in which Anne Baker convincingly discusses the lasting effect of geography primers such as *Peter Parley*'s on how generations of Americans growing up during the nineteenth century perceived their national identity, racial differences, and the world. Baker, *Heartless Immensity*, chap. 7.

33. Goodrich, *Peter Parley's Method of Telling about Geography*, 75.

34. Goodrich, *Peter Parley's Method of Telling about Geography*, 74, 72.

35. Goodrich, *Peter Parley's Method of Telling about Geography*, 71, 57, 41.

36. Goodrich, *Peter Parley's Method of Telling about Geography*, 91.

37. In antebellum but also later geography schoolbooks, teaching about astronomy and its accompanying religious lessons were a favorite subject. Indeed, the section of school geography books called "astronomical geography" pursued a conflation of religious and scientific objectives. In 1784, Jedidiah Morse reminded young Americans—and especially their parents and teachers—that "a complete knowledge of Geography cannot be obtained without some acquaintance with Astronomy.... Astronomy treats the heavenly bodies and explains their motions, times, distances and magnitudes. The regularity and beauty of these, and the harmonious order in which they move, show that their Creator and Preserver possesses infinite wisdom and power." Morse, *American Universal Geography*, 1. The section on astronomical geography with which Morse's textbooks started did not include any illustrations. In a similar vein, Samuel G. Goodrich's geography book made an unsophisticated, inaccurate, and cursory reference to astronomical geography in a strictly Christian tone: "It is now nearly six thousand years since God created this world on which we

live. He made it, and swung it in the air, and ever since he has kept it moving with the other planets through the heavens." Goodrich, *Peter Parley's Method of Telling about Geography*, 79.

38. For a study of astronomical geography, which was included in geography textbooks before 1850, see Vining, "Astronomical Geography," 30–40.

39. Goodrich, *Peter Parley's Method of Telling about Geography*, 79.

40. Niles, *Complete Geography*. For a late nineteenth-century map showing the distribution of "races of men," see figure 1–7.

41. Swinton, *Grammar-School Geography: Physical, Political, and Commercial*, 108–9.

42. This was, of course, not true of all geographies published by this time. Edwin J. Houston's *Elements of Physical Geography* (in print between 1891 and 1905), for instance, featured a map of the world with the Americas to the left of the projection.

43. Roddy, *Complete Geography*, 140–41.

44. Stratton, *Education for Empire*, 17.

45. For a compelling account of the changes in American school curricula in a historical context, see Shepherd and Ragan, *Modern Elementary Curriculum*. For a comparably thorough study of American educational reforms since the 1890s, see Kliebard, *Struggle for the American Curriculum*.

46. Quoted in Bourne, *Teaching of History and Civics*, 59.

47. Bourne, *Teaching of History and Civics*, 59.

48. Schulten, *Geographical Imagination in America*, 104.

49. Hinsdale, "Questions in Geography and History," 8–9.

50. Hinsdale, "Questions in Geography and History," 8–9.

51. Hinsdale, "Questions in Geography and History," 8–9.

52. Gordy and Twitchell, *Pathfinder in American History*, 20–21.

53. Gordy and Twitchell, *Pathfinder in American History*, 23.

54. Koelsch, "Academic Geography, American Style," 248.

55. According to John Seiler Brubacher and Willis Rudy, the number of American students who took geography at the college level increased ten times between 1900 and 1948. See Brubacher and Rudy, *Higher Education in Transition*; quoted in Murphy, "Geography's Place in Higher Education," 3. However, the growing interest in academic geography at the turn of the twentieth century seems not to have been a long-lived phenomenon. In fact, during and shortly after the end of World War II, some of the departments of geography at leading American universities closed down, starting with Harvard's in 1948 and followed by those at Columbia, Michigan, Pennsylvania, Stanford, Virginia, and Yale. For an examination of the rise and fall of geography as an academic discipline in response to the later rise of disciplines such as the political sciences and politics, see Steinmetz, "Geopolitics"; Murphy, "Geography's Place in Higher Education."

56. Brooks, *Index to the Journal of Geography*, iii.

57. Dexter, *History of Education in the United States*, 162.

58. Dexter, *History of Education in the United States*, 162.

59. The geography primers that Dodge authored or coauthored, such as *Principles of Geography, Home Geography and World Relations*, and *Our Neighbors across the Seas*, were all published by Rand McNally, the prestigious producer of maps, atlases, and geography books. In these volumes, the author(s) always placed the United States in a larger planetary

web of interdependence. At the same time, Dodge promoted a comparative approach to teaching geography and advocated world geography in his books as well as in the journal he edited. The journal was discontinued in 1901 but was soon replaced by the *Journal of Geography*, which Dodge and Edward M. Lehnerts, professor of geography at State Normal School, Minnesota, coedited.

60. Written by the American explorer-turned-journalist Thomas Wallace Knox (1835–96) and published by Harper and Brothers, The Boy Travellers series included several titles, as follows: *The Boy Travellers in the Far East* (in five parts, published between 1879 and 1883); *The Boy Travellers in South America* (first published in 1885); *The Boy Travellers in the Russian Empire* (first published in 1886); *The Boy Travellers on the Congo* (1887); *The Boy Travellers in Australasia* (first published in 1888); *The Boy Travellers in Mexico* (first published in 1889); *The Boy Travellers in Great Britain and Ireland* (first published in 1890); *The Boy Travellers in Northern Europe* (first published in 1891); *The Boy Travellers in Central Europe* (first published in 1892); *The Boy Travellers in Southern Europe* (first published in 1893); and *The Boy Travellers in the Levant* (first published in 1894).

61. Both the term "home geography" and the pedagogic baggage associated with it were in use well into the twentieth century. Mindy Spearman believes that while this approach to teaching geography was in vogue until the 1920s, the term was still commonly recognized by geographers as late as the 1960s. Spearman, "Race in Elementary Geography Textbooks," 115.

62. Davis, "Home Geography," 2.

63. Davis, "Home Geography," 6.

64. Davis, "Home Geography," 6.

65. An astute outdoor observer and a firm believer in the basics of evolution in geographical interpretations, Davis was celebrated in his biography by Reginald A. Daly as "an apostle bringing to them [geographers and geologists] the gospel of method in research and method in the presentation of the results of research." Daly, *Biographical Memoir of William Morris Davis*, 263.

66. R. E. Dodge, *Home Geography and World Relations*.

67. Dodge, *Home Geography and World Relations*, 3.

68. Dodge and Kirchwey, *Teaching of Geography in Elementary Schools*, 3.

69. Dodge and Kirchwey, *Teaching of Geography in Elementary Schools*, 31.

70. Dodge and Kirchwey, *Teaching of Geography in Elementary Schools*, 40–41.

71. Dodge and Kirchwey, *Teaching of Geography in Elementary Schools*, 41–42.

72. Kaplan, "Manifest Domesticity," 588. Kaplan's reference here is to the rhetoric of expansionism in the 1840s and the 1850s as public figures such as Sara Josepha Hale, Catherine Beecher, John L. O'Sullivan, and Horace Mann interwove the sentimental logic and the language of domesticity with the aggressive logic and the language of the so-called manifest destiny.

73. See chapter 4, in which I expand on this binary in the study of children's letters.

74. I return to this question in chapters 3 and 4, in which I examine the ways children made sense of the world based on this model.

75. Hubbard, "Practice School Course in Geography," 393.

76. Hubbard, "Influence of the Presence, Discovery and Distribution of Precious Metals."

77. Brigham, "Physical Geography in Secondary Schools," 924.

78. Brigham, "Physical Geography in Secondary Schools," 924.

79. For an understanding of Blache's contributions to geography, his debt to German geographer Friedrich Ratzel, and the European roots of his thought, see Chorley and Haggett, *Frontiers in Geographical Teaching*; Malpas, *Heidegger and the Thinking of Place*; Febvre et al., *Geographical Introduction to History*.

80. Beyond a doubt, certain of those localities were sure to be more similar and less distant to the American home than others. After all, "the world" that modern American geography claimed to be at the center of its focus was cast in different lights, with different areas of emphasis, at different times.

81. In Rebekah Sheldon's words, "The child stands in the place of the species and coordinates its transit into the future." Sheldon, *Child to Come*, vii.

82. I return to this topic throughout the remainder of this volume, especially in discussions on children's relationship to renaming the colonized world as an extension of home geography.

83. *Harper's School Geography, with Maps and Illustrations*, 1. Earlier versions of the same book had been copyrighted as early as in 1844 under the title *A System of Geography for the Use of Schools: Illustrated with More Than Fifty Cerographic Maps and Numerous Wood-Cut Engravings*.

84. Pratt, *Guyot Geographical Reader and Primer*, 1. The reader and primer were first published by Charles Scribner's Sons in 1882 (the quotation is from this first edition) and were later copyrighted by the American Book Company in 1898.

85. Pratt, *Guyot Geographical Reader and Primer*, 3.

86. Pratt, *Guyot Geographical Reader and Primer*, 184.

87. Pratt, *Guyot Geographical Reader and Primer*, 208.

88. For a discussion of the racist undertones of polygenism as a tool deployed to justify slavery in the United States and of its proponents and opponents, see Wolpoff and Caspari, *Race and Human Evolution*. For a fine-grained historical analysis of prevalent discourses on race and racial differences from before the age of exploration to the present and of the changing bases of racism from mythical to biological and now ideological, see Graves, *Emperor's New Clothes*.

89. C. F. King, *Picturesque Geographical Reader*.

90. G. S. Hall, "Child Study in Summer Schools," 333–36.

91. Jacob Abbott (1803–79) was the first of many nineteenth-century American authors to turn to the burgeoning children's book market of the time, producing a travel literature series that would send off the American boy Rollo (sometimes accompanied by his sister) on travel adventures back in Europe and beyond. For a thorough examination of these serial works and their content, see Nesmith, "Young Americans Abroad."

92. C. F. King, *Picturesque Geographical Reader*, 4.

93. C. F. King, *Picturesque Geographical Reader*, 6.

94. For an examination of American missionary work during the long nineteenth century, see, among others, Tyrrell, *Reforming the World*; Case, *Unpredictable Gospel*. Sharkey, *American Evangelicals in Egypt*, is another illuminating example that focuses on U.S. missionary activities in Egypt. A further illuminating case study is offered in Baker, *Revival and Awakening*. In *Dependent States*, Sánchez-Eppler explores the complex relationship among American home life, parenting, and missionary work as the building blocks of a globalizing U.S. empire.

95. According to Andrew C. Godley, foreign investment in the American economy was unusually high from the very beginning. Starting at $70 million in 1803, it reached $3 billion in 1900. Godley, "Foreign Investment," 487. For a detailed account of the economic development of the United States in the long nineteenth century, see Davis and Cull, *Capital Movements, Markets, and Growth*.

96. Gallman and Engerman, *Cambridge Economic History of the United States*, 688.

97. According to Richard Phillips, "sketchy" or "sketch" maps are informal hand-drawn maps, usually unicolored, that typically lack elaborate mathematical scaling. Inculcating their viewers with "an alluring and ambiguous geographical suggestion," sketch maps are further void of common cartographic symbols and "scientific pretensions." Phillips, *Mapping Men and Empire*, ix.

98. Said, *Culture and Imperialism*, 7.

99. For illuminating analyses of cartography in colonial agendas, see Said, *Culture and Imperialism*; Anderson, *Imagined Communities*.

100. Anderson, *Imagined Communities*, 173–74.

101. Anderson, *Imagined Communities*, 174.

102. In addition to Phillips's *Mapping Men and Empire*, insightful discussions are made about geography as a masculine practice in Sinha, *Colonial Masculinity*.

103. On the influence of the works by prominent German geographer Friedrich Ratzel's disciple Ellen Semple (1863–1932) at the turn of the century, see Keighren, *Bringing Geography to Book*.

104. Schulten, *Geographical Imagination in America*, 3.

105. Du Bois, *Du Bois on Education*, 205.

106. Dodge and Kirchwey, *Teaching of Geography in Elementary Schools*, 40.

Chapter Two

1. For a thorough examination of the "Nellie Bly phenomenon" and the publicity surrounding her trip around the world, see Wong, "Around the World and across the Board."

2. Chaplin, *Round about the Earth*, 221.

3. For an account of around-the-world journeys made by Americans in a wider context of global circumnavigation over the centuries, the public craze over the media-covered trips at the end of the nineteenth century, and the occasional fraudulent cases, see Chaplin, *Round about the Earth*, 221–54.

4. Braden, "'The Family That Plays Together,'" 145–46. See also Shrock, *The Gilded Age*, 151–52.

5. "Industrial Age, Recreational Life," 604.

6. Twain, *Innocents Abroad*.

7. Inspired by the Spanish-American War of 1898, The Game of War at Sea, or Don't Give Up the Ship!, sends its young players on a quest to find out whether "the United States can beat off or destroy a hostile fleet in American waters." For a list of "map games" produced by McLoughlin Brothers in 1899, see "$1.00 Board Games" in *McLoughlin Bros.' Catalogue*, 94–100.

8. Braden, "The Family That Plays Together," 121–25.

9. For an extensive collection of firsthand historical documents and records on the firm, see "McLoughlin Brothers Papers"; "McLoughlin Bros. Collection." See also the catalog of

an eponymous exhibition of the picture books published by the company in Hewes et al., *Radiant with Color and Art.*

10. Bernstein, "Children's Books, Dolls, and the Performance of Race." Bernstein deploys a similar analytical method in *Racial Innocence*, 8.

11. In this sense, childhood had joined other demographics of American citizenry in departing from religious moralization toward popularization and politicization, while turn-of-the-century mass-produced American toys had started carrying the label "amusing and instructive." See Weinstein, *Once Upon a Time*, 9.

12. See Hannas, *The Jigsaw Book*; Williams, *Jigsaw Puzzles*; Shefrin, *Neatly Dissected*. See also Shefrin, *Such Constant Affectionate Care*. For a comprehensive take on the age of reason and the significance of geography in, and at the service of, the Enlightenment project, see Withers, *Placing the Enlightenment*. See also Bowen, *Empiricism and Geographical Thought*.

13. For further details on this, see Shefrin, *Neatly Dissected*, 7–24.

14. Hannas, *The Jigsaw Book*, 9–12.

15. Hannas, *The Jigsaw Book*, 9–12.

16. A mapmaker with a knack for business, Spilsbury had two sets of prices for his generally expensive dissected maps—one with and one without the sea. Hodgkiss, *Discovering Antique Maps*, 101.

17. Examples include picture puzzles that depicted scenes from the Old Testament, including the lives of prophets. Hannas, *The Jigsaw Book*, 11.

18. Anne Williams's illuminating study of puzzles indicates that the so-called metamorphosis puzzles made it possible for puzzle solvers to put them together in odd, comical orders and come up with disfigured images in the end. But this was only possible if the puzzles' pieces were of equal size—a novelty that enjoyed unprecedented popularity at the end of the nineteenth and early twentieth centuries and came to be an obsession of both children and adults, especially during times of crisis, such as the Great Depression. Williams, *Jigsaw Puzzles*, 5–8.

19. Williams, *Jigsaw Puzzles* 5; Hannas, *The Jigsaw Book*, 11.

20. During the time when Germany was leading toy manufacturing worldwide, German puzzles—usually labeled in three or four different languages—were exported to other European countries as well as to the United States. According to Hannas, a great majority of nineteenth-century puzzles found in Scandinavia, for example, were manufactured in Germany. Hannas, *The Jigsaw Book*, 15.

21. For further details on early American dissected maps, see Hannas, *The Jigsaw Book*, 15–24.

22. For a careful discussion of U.S.-manufactured map puzzles as object lessons between 1800 and 1860, see Brückner, *Geographic Revolution*, chap. 3; Brückner, *Social Life of Maps*, chap. 8.

23. *Confidential Price List and Telegraphy Code.*

24. Texas Institution for the Blind, *Annual Report*, 10–11.

25. For further details, see Marsh and Millard, *Literacy and Popular Culture*, 53–54. Toys, Marsh and Millard assert, exist in "an intertextual universe" (54).

26. David Rumsey Historical Map Collection.

27. Williams, *Jigsaw Puzzles*, 3.

28. According to the U.S. census, the percentage of the U.S. population living in a city of 2,500 or more had increased from 5.1 in 1790 to 35.1 by 1890 (United States Census Bureau, "Increasing Urbanization"). For a historical examination of American children's toys and games, see Chudacoff, *Children at Play*. Chudacoff shows throughout the book that no single definition of what play is can be agreed upon across disciplinary boundaries.

29. Sutton-Smith, "Does Play Prepare for the Future?," 141–42.

30. Harvey, *Condition of Postmodernity*, 220.

31. Mayar, "Verbs of Violence."

32. Brody, *Visualizing American Empire*, 97.

33. Schulten, *Geographical Imagination in America*, 29.

34. Anderson, *Imagined Communities*, 179. For a fine-grained examination of early-American practices in color-coding maps in both national and imperial contexts, see Brückner, *Social Life of Maps*, esp. chaps. 3, 7.

35. Anderson, *Imagined Communities*, 179.

36. Though preceding the time when jigsaw puzzles made their way into the popular imagination of youths on both sides of the Atlantic, such a jigsaw-like power relationship was already present in aristocratic school curricula and imperial cartographers' workshops alike well before Madame Beaumont and John Spilsbury used scissors and handsaws in order to dissect them for wider public consumption. For further details on U.S. imperialism and geopolitics in school geography books, see chapter 1.

37. Notably, the absence of the family's father, who might be at work, suggests a cultural ideal in which the middle-aged male figure has his job and his social life outside the home.

38. Malte-Brun, *Précis de la géographie universelle*, quoted in Mitchell, *Mitchell's Geographical Reader*. Summing up the discussions presented in chapter 1 and here, I would like to emphasize that the messages coded by American schoolbooks and children's toys were, at least in a great many instances, remarkably similar.

39. For an understanding of the fierce international competition for China's market, including the United States' advocacy for an open-door policy in interimperial dealings with China, see, among others, Elleman, *International Competition in China*; Moore, *Defining and Defending the Open Door Policy*.

40. There is no doubt that jigsaw puzzles could be found in the American market even before the Civil War. In her book-length study of jigsaw puzzles in the United States, Williams reports that by the mid-nineteenth century, British, German, and French puzzle makers had already started to produce dissected maps exclusively designed for the American market. In addition to McLaughlin Brothers and S. B. Ives, Williams also includes a long list of American puzzle makers, including Samuel L. Hill of Brooklyn, V. S. W. Parkhurst of Providence, and Thomas Wagner of Philadelphia, who had already begun to produce puzzles for the American market in the antebellum era. Williams, *Jigsaw Puzzles*, 6–7.

41. Paul Reddin's *Wild West Shows* delivers an exhaustive study on the history, formation, and egregiously racist features of Wild West shows over the course of the nineteenth century. Beginning with George Catlin from the 1830s to the 1860s, the book records and analyzes the show's popularity in Europe. For fine-grained examinations of the image of "the Indian" in Wild West shows during the nineteenth and twentieth centuries and on both sides of the Atlantic, see Moses, *Wild West Shows*; McNenly, *Native Performers in Wild West Shows*. Moreover, in *Performing the American Frontier*, Roger A. Hall provides a greater

turn-of-the-century tradition, which gave rise to a wide range of public performances such as Wild West shows.

42. Milton Bradley, *Catalogue of Games, Sectional Pictures, Toys, Puzzles, Blocks and Novelties*.

43. In general, the second half of the nineteenth century, and especially the 1880s and 1890s, witnessed a boom in consumer culture across the United States, including the proliferation of affordable child-specific goods from books and toys to clothing and furniture. By this time, children's consumer culture had taken a long path from folk to popular culture. Over the course of two centuries, repeated encounters of the white colonizers with Native Americans—as a result of which white Americans had begun making and playing with corn husk dolls, clay marbles, and pick-up sticks—had been gradually exchanged for a fiercely profit-oriented market that manufactured popular toys and ABCs in response to the changing tides of the time. A locus classicus for the study of nineteenth-century children's entertainment through the examination of children's materials produced by toy manufacturers like Milton Bradley and McLoughlin Brothers is Weinstein's *Once Upon a Time*, a volume that highlights the indivisibility of personal interest, nostalgia for childhood memories, collectors' thirst for material culture, and archival adventurism as prime features of children's archives.

44. Epple and Schaser use the term "gender order" in reference to the naturalized gender hierarchy in popular and academic historiography of the West before gender studies entered the equation to question taken-for-granted gender constructs and to disrupt the balance in favor of new perspectives and agencies in history writing. Epple and Schaser, *Gendering Historiography*, 7–9. For further discussions of the various applications of the term "gender order" in historiography in the past two decades, see, among others, Jordanova, "Gender and the Historiography of Science"; Berger and Lorenz, *Contested Nation*; Harris, "Sex on the Margins"; Haggis, "Gendering Colonialism or Colonising Gender?"

45. Sánchez-Eppler, *Dependent States*, 186–220.

46. In her fascinating study of the early nineteenth-century conflations between the language of expansion with the language of female domesticity, Amy Kaplan offers "manifest domesticity" as a frame through which to ponder the contradictions at the heart of the simultaneously domestic, national, and global project of empire. Kaplan, "Manifest Domesticity."

47. For an illuminating discussion of intersectionality between race and gender against the backdrop of national politics, see Grabham, *Intersectionality and Beyond*; Collins, "It's All in the Family."

48. For further discussions on the gendered nature of manifest destiny and the role of women at the forefront of imperial projects of the time, see A. S. Greenberg, *Manifest Manhood and the Antebellum American Empire*. Richard Phillips's *Mapping Men and Empire*, on the other hand, discusses the European roots of masculinity in the imperial geographical imagination, asserting that while pushing women aside in dominant discourses during the course of the nineteenth century, "geographical fantasy belong[ed] to the man" (5).

49. *Confidential Price List and Telegraphy Code*, 127.

50. For Roosevelt's account of the incident, which appeared in serial form in *Scribner's Magazine*, see Roosevelt, *The Rough Riders*, 67–95.

51. For Presley's full response, see Washington, Wood, and Williams, *New Negro for a New Century*. On the formation of the Rough Riders regiment in 1898 and how Theodore

Roosevelt's role as its second commander in the Battle of San Juan Heights facilitated his later rise to presidency, see Samuels and Samuels, *Teddy Roosevelt at San Juan.*

52. Brandt, "Writing Political History after the 'Iconic Turn,'" 356.

53. For an insightful examination of the role of the United States in the international community as an arbiter of peace, see Witt, *Patriots and Cosmopolitans.*

54. Witt, *Patriots and Cosmopolitans,* 178–80.

55. "Peace Work at the American Institute of Instruction," 169.

56. "Peace Work at the American Institute of Instruction," 169. The meeting was held in Boston as part of the annual meeting of the American Institute for Instruction from July 9 to July 12, 1906. The report mentions that the meeting was a huge success, as the organizers had to give up the room where the meeting was originally scheduled in order to hold it in a larger hall. For a history of the institute from its inception in August 1830 until its sixty-sixth annual meeting in 1906, see "American Institute of Instruction, 1830 to 1906."

57. Deleuze and Guattari, *A Thousand Plateaus: Rhizomes,* 13.

58. Deleuze and Guattari, *A Thousand Plateaus: Rhizomes,* 11–14.

59. Wood and Fels, *The Power of Maps,* 5.

60. Schulten, *Geographical Imagination in America,* 3.

61. Sánchez-Eppler, *Dependent States,* 60.

62. As Forman-Brunell asserts in the case of doll play, for example, "Abuse of dolls at the hands of their owners alerts us that adult prescriptions for proper play were often not what girls had in mind." Forman-Brunell, *Made to Play House,* 27.

Chapter Three

1. As the discussions in this and the following chapters confirm, *Harper's Young People* and *St. Nicholas* were by no means unidirectional mouthpieces for professional adult writers to convey American values, including class-conscious social norms and gender-specific cultural standards, to their young audiences. Rather, they presented strictly edited yet interactive spaces where intergenerational conversation (between professional writers and editors, child readers and writers, and young adult readers, writers, and ghost writers) was welcomed. Gannon, for instance, comments on the interactions between different generations in *St. Nicholas* as "an extended and multidirectional conversation." Referring to the role readers' responses to the magazine played in influencing the editorial policies established by Mary Mapes Dodge during the long years of her charismatic editorship, Gannon offers a fresh understanding of such interactions as a "space in between," in which "differing—sometimes age-specific—visions of child-adult relations could be figured, tested, and vigorously discussed." Gannon, "'Best Magazine for Children of All Ages,'" 153.

2. *Harper's Young People* 1, no. 5 (December 2, 1879).

3. *Harper's Young People* 1, no. 9 (December 30, 1879).

4. *Harper's Young People* 1, no. 11 (January 13, 1880).

5. *Harper's Young People* 1, no. 16 (February 17, 1880).

6. *Harper's Young People* 1, no. 15 (February 10, 1880).

7. Bach, *Colonial Transformations,* 69.

8. Said, *Orientalism*, 162.

9. Danesi, *Puzzle Instinct*, 3.

10. For an overview of Sam Loyd's inventions as well as his involvement in copyright scandals, see the introduction to Gardner, *Mathematical Puzzles of Sam Loyd*; Slocum, "Sam Loyd's Most Successful Hoax."

11. Equally striking is the involvement of L. Frank Baum, author of *The Wonderful Wizard of Oz* (1900), in the elections. Unlike Loyd, Baum supported the Democratic Party's candidate William Jennings Bryan. For a compelling reading of *The Wonderful Wizard of Oz* as a parable of the 1896 presidential elections, see Littlefield, "The Wizard of Oz."

12. See introduction to Gardner, *Mathematical Puzzles of Sam Loyd*. The solution to the puzzle can be found in Danesi, *The Puzzle Instinct*, 92–93.

13. McKinley's wife was chronically ill, so the McKinleys could not travel across the country during the campaign period.

14. The puzzle's mystery lies, rather simply, in its use of optical illusion and the way the cut-out bits of the Chinese figures' bodies both *almost* match and *almost* overlap. When the puzzle's inner disk (the earth) is turned counterclockwise, one of the thirteen Chinese figures "gets off the earth."

15. An ethnic slur, "jap" was used in vernacular English in the nineteenth century mostly as a derogatory term in reference to people of Japanese descent. The term gained exceedingly offensive racial connotations in and after the bombing of Pearl Harbor.

16. Danesi, *The Puzzle Instinct*, 92.

17. Costello, *Greatest Puzzles of All Time*, 108.

18. For a careful examination of racialization and population management in late-century U.S. policies toward Chinese and Filipino immigrants as "enumerative strategies" of the empire, see Hsu, "Body Counts and Comparative Imperialism."

19. Some of the reasons behind this renewed imagined crisis in the United States at the turn of the twentieth century and the political repercussions of tackling it are discussed in Hoganson's in-depth work, *Fighting for American Manhood*. For broader examinations of the crisis of masculinity as a recurrent historical discourse and a marker of gendered national crises across the globe, see, among others, Bederman, *Manliness and Civilization*; Hooper, *Manly States*; A. S. Greenberg, *Manifest Manhood*; Haschemi Yekani, *Privilege of Crisis*; Wendt and Andersen, *Masculinities and the Nation in the Modern World*. A more recent conversation in direct reference to the exigencies of empire is made in Honeck, *Our Frontier Is the World*.

20. Danesi, *The Puzzle Instinct*, 208.

21. See, for instance, Gardner, *Mathematical Puzzles of Sam Loyd*; Gardner, *More Mathematical Puzzles of Sam Loyd*. Further studies on Sam Loyd include Pickard and Loyd, *Puzzle King*, as well as Sam Loyd's own collection of puzzles and tricks, Loyd, *Sam Loyd's Cyclopedia of 5,000 Puzzles*.

22. As noted earlier, for more than a decade, the magazine regularly printed a number of children's puzzles, including geographical puzzles, until they stopped the column on March 3, 1891.

23. A scene of colonial contests and conquests between Arabs, Indians, and Europeans on a small scale over the centuries, the thinly populated Alphonse Isle is now a luxury resort and a tourist attraction as one of the many islands in the Seychelles Islands nation. For general information on the island, see Alphonse Island, "For the Ultimate Getaway." For a

brief account of the island's history, see Alphonse Island, "Brief History of Alphonse Island and the Seychelles."

24. Harking back to the "Norfolk Tragedy," a sixteenth-century tale in which orphaned children are left in the woods to die, and in line with late-nineteenth-century attempts by American parents to domesticate children and child play, a strikingly high percentage of the story puzzles I study here made ambivalent, at times contradictory, comments on whether it was safe for children to play outside and without adult supervision. Reading these "babes in the wood" tales next to the narratives that placed the nation in the world, I find it of immense significance to note the tensions at work as the nation remapped the westward expansion across the Pacific and into the Caribbean.

25. Roughly put, homonyms are words that share the same spelling (homographs), share the same pronunciation (homophones), or both. For example, while "saw" (past tense of the verb "see") and "saw" (tool to cut wood) are both homographs and homophones, "see" and "sea" are only homophones. The relationship between homonyms is called homonymy (that is, having the same name).

26. Brückner, *Early American Geographies*, 83. In turn, postcolonial scholarship has closely engaged with renaming in colonial as well as postcolonial and emancipation contexts. Postcolonial scholars view renaming as part and parcel of the mapping side of colonization and a means of control and mastery, exclusion, and dispossession. In this frame, renaming has been viewed as a (post-)colonial means to distort, and later to restore, natives' kinship with their surroundings, their culture and language, and indeed their bodies. See, among others, Ashcroft, Griffiths, and Tiffin, *Postcolonial Studies: The Key Concepts*, 39; Calloway, *New Worlds for All*, 1–7.

27. For multifaceted discussions on child education, children's literature, and ecocriticism in the English-speaking world, see Dobrin and Kidd, *Wild Things*. For a more recent, equally remarkable work that engages with representations of nature and ecocriticism in children's literature and culture in northern Europe, see Goga et al., *Ecocritical Perspectives on Children's Texts and Cultures*.

28. For similarities in what mattered to children when they wrote letters to magazines, see chapter 4.

29. For details about Henry M. Stanley's expeditions and the controversies in which he was involved, see his autobiography, Stanley and Stanley, *Autobiography of Sir Henry Morton Stanley*. See also Stanley, *Through the Dark Continent*; Jeal, *Stanley*.

30. Knox and Stanley, *Boy Travellers on the Congo*.

31. Knox and Stanley, *Boy Travellers on the Congo*, 13.

32. The Stanley Falls were later renamed once more to—and are currently called—Boyoma Falls. Located in the Democratic Republic of Congo, Boyoma Falls includes seven short falls spanning a large area where the Lualaba and Congo Rivers meet.

33. Pratt, "Science, Planetary Consciousness, Interiors."

34. For yet another instance of renaming in the case of the Apache children, see chapter 4. For further evidence from various historical settings, see, among others, Ntamushobora, *Education for Holistic Transformation in Africa*, esp. chap. 2; Bailey, *Race and Redemption in Puritan New England*, esp. chap. 4; C. Phillips, *Freedom's Port*, esp. chap. 4. There, Phillips discusses two instances of renaming: one that robbed first generations of slaves of their African identities when they were enslaved, and the other that, at least symbolically, restored their freedom and identity when they changed their names once they were freed. Furthermore, K. S. Green-

berg, *Nat Turner*, especially chapter 1, refers to given names as having individual, communal, and even territorial qualities to them. Finally, see Clark, Hercus, and Kostanski, *Indigenous and Minority Place Names*, a volume that provides a valuable linguistic and ethnographic analysis of place names, hegemonic historical forces, and Indigenous identity among various colonized tribes in regions as diverse as Australia, Africa, and North America.

35. Harvey, *Condition of Postmodernity*, 220.

36. For a fine-grained analysis of mapping as an organizational tool and its various features as a bridge between reality and the patterns behind it, see Weick, "Cartographic Myths in Organizations," 1–10.

37. The Swiss psychologist Jean Piaget (1896–1980) founded his analysis of young children's understanding of the world on the minimal assumption that children know a place simply if they can name it. His numerous studies have come under criticism by scholars who either found his assumptions too basic or raised concerns about their reproducibility in other settings. Indeed, regardless of the striking breadth of studies he conducted throughout his career, some of his conclusions concerning the complex and mutable particularities of acquiring and retaining geographic knowledge are too offhand to be entirely credible. For an overview of the wide range of studies he conducted on children's cognition, see Piaget, *Child's Conception of the World*; Bringuier and Piaget, *Conversations with Jean Piaget*; Piaget and Weil, "Development in Children of the Idea of the Homeland"; Piaget and Inhelder, *Child's Conception of Space*. For a fine roundup of the critique of his work, see Wiegand, *Places in the Primary School*, 38–47.

38. See Karl Weick's arguments to this effect in his discussion of cartography in relation to territory. Weick, "Cartographic Myths in Organizations," 2.

39. Crain, *Story of A*, 56.

40. Biddick, "The ABC of Ptolemy," 268–94.

41. Biddick, "The ABC of Ptolemy," 268–94.

42. The politics of making literacy an exclusive privilege and denying public access to reading and writing as a means to practice power over the public by the church should not mask the fact that letters of the alphabet have endured far more complex processes of negation and change over the centuries than mere power politics, which were involved in the interpretation of the Latin Bible during the Middle Ages. Due to their ability to convey multiple and opposing meanings, letters of the alphabet have historically been regarded by those who knew how to use them as enabling tools that make unexpected utterances and associations possible—utterances and associations that were not necessarily in the interest of the powerful. Therefore, letters of the alphabet were long considered potentially dangerous, hence the church's preference to keep literacy in its own hands.

43. One of the oldest forms of writing that is still in use today, Hanzi characters are logograms used to write Chinese and several other East Asian languages. The characters have a syllabic basis, meaning that unlike letters of the alphabet that represent individual sounds, each Hanzi logogram corresponds with a syllable (if not more). Hanzi characters are adopted as the written form in various East Asian languages, such as Japanese and Korean.

44. While most probably a sign of Sam Loyd's backhanded indifference toward this particular instance, it is too convenient, if not outright perilous, to reduce racism to laziness. Even in its least harmful forms, racism as a public projection of hatred, even as a passive collective sentiment, is more than merely lazy or uninterested.

45. For an exemplary examination of ABC books at the service of socializing American children into their future adult roles, see Crain, *Story of A*.
46. Crain, *Story of A*, 64.
47. Crain, *Story of A*, 64.
48. Crain, *Story of A*, 91.
49. Crain, *Story of A*, 96–99.
50. Piaget and Weil, "Development in Children of the Idea of the Homeland," 561–78. As noted before, criticized by later generations of scholars in the fields of child psychology and education for their lack of clear-cut definitions and the impossibility of repeating them in other settings, Piaget's studies at the University of Geneva are still worth returning to, especially since they put together ideas about childhood, nationalism, internationalism, and spatiality that had long been talked about but had hardly ever been examined methodologically.
51. Piaget and Weil, "Development in Children of the Idea of the Homeland," 563.
52. Piaget and Weil, "Development in Children of the Idea of the Homeland," 564–65.
53. For a detailed discussion of this dual-world model, see chapter 1.
54. It is particularly interesting to notice when we compare this location distribution pattern with the findings of Matthew Wilkens's groundbreaking geolocational text mining of "the imaginative landscape of American fiction" in the nineteenth century. As the results of Wilkens's data mining of works of fiction listed in the Wright American Fiction Collection confirm, on aggregate, nineteenth-century American fiction has been far less regional and far more "diversely outward looking" and international than literary scholars had long believed it to be. For a fascinating account of Wilkens's digital humanities project on nineteenth-century American fiction, see Wilkens, "Canons, Close Reading, and the Evolution of Method."
55. For a compelling trail-blazing study of letters of the alphabet and their shifting trifold functions in child literacy—internalization, representation, and ordering—since the seventeenth century, see Crain, *Story of A*, chap 2.
56. Frankfort, as spelled by A. Lloyd Cooper, had little to do with the German city Frankfurt. A site of early conflicts between Native Americans and the white settlers, Frankfort was an American town that became the capital of the state of Kentucky in 1792.
57. According to Wiegand's findings, Zimbabwe was mentioned by several British children ages seven and eleven for two reasons: first, that children "just liked saying it"; and second, that Zimbabwe reminded British children of their favourite Liverpool F. C.'s goalkeeper, who was from Zimbabwe—both facts underlining a wider cognitive picture from which children picked geographical names. Wiegand, *Places in the Primary School*, 68.
58. For illuminating details on the Anglo-German conflict in Africa, including the Anglo-German Agreement of 1890, see Gillard, "Salisbury's African Policy and the Heligoland Offer of 1890"; Sanderson, "Anglo-German Agreement of 1890"; Holborn, *History of Modern Germany*.
59. Striking is the fact that Zanesville, Ohio, is surrounded by European place names (from Dresden, Hanover, and Vienna to Cambridge, Warsaw, and Athens), all within a few miles distance.
60. Burnett made a career first and foremost as a writer for children. Her debut novel, *Little Lord Fauntleroy*, was initially serialized in *St. Nicholas* in 1885 and 1886, and

its theatrical adaptions in the late 1880s and early 1890s were met with huge success across urban America. She wrote her other most well-received novel, *The Secret Garden*, in 1911 in London. Burnett died on Long Island on October 29, 1924. For an illustrative account of Burnett's life and career, see Gerzina, *Frances Hodgson Burnett*.

61. Within a few years after Marie's puzzle was printed in *Harper's Young People*, Frances Hodgson got divorced from her first husband, Swan Burnett, thus leaving behind both the United States and the letter in her name that stood for it.

62. Two years before *Little Lord Fauntleroy* was released in serial form in *St. Nicholas*, Frances Hodgson Burnett published *The One I Knew Best of All: A Memory of the Mind of a Child*, in which she wrote about her own childhood, including an account of her staging of *Uncle Tom's Cabin*, during which she cast the Black doll she owned as Uncle Tom and whipped him. For a thoughtful discussion of Burnett's recollections of racial violence couched as "innocent" child play and the influence her writing about it had on the young fans of *Little Lord Fauntleroy* and *Uncle Tom's Cabin*, see Bernstein, *Racial Innocence*, esp. chap. 2.

63. The topsy-turvy doll was believed to have first been made by enslaved African American women in the plantations of the South. This peculiarly cross-racial doll has been examined by a great number of historians of childhood who have read it in terms of racial uncertainty and identity reversal (Shirley C. Samuels); its conflation of domesticity, play, and (a)sexuality (Karen Sánchez-Eppler); the material it was made of and how that changes child play (Robin Bernstein); and the ways it codes sexual violence against (Black) women in a plaything for (white) girls (Kimberly Wallace-Sanders). Here, reading about Burnett's racial violence to Black dolls against the backdrop of Marie Buchanan's attempt to make the spelling of the republic possible with letters of her name, I view the rag doll's skirt as the most intimate and gendered of spaces to be found in the nineteenth-century American home. Samuels, *Romances of the Republic*, 113; Sánchez-Eppler, *Touching Liberty*, 133–41; Bernstein, *Racial Innocence*, 81–91; Wallace-Sanders, *Mammy: A Century of Race, Gender, and Southern Memory*, 32–35.

64. In *The Story of A*, Crain deploys the phrase "republic of alphabet" in reference to the nationalistic function that the alphabet began to have in the United States since the turn of the eighteenth century. Crain, *Story of A*, 56.

65. This was part of puzzle No. 6, under the heading "Pied Cities," printed in the March 4, 1890, issue, with the correct answers to it printed in the March 25, 1890, issue.

66. For more on the significance of Hong Kong to the British Empire and the United States, see Ágoston and Masters, *Encyclopedia of the Ottoman Empire*; *Commercial Relations of the United States with Foreign Countries*. For a thorough understanding of the complex interimperial rivalries at work in nineteenth-century Hong Kong, see Share, *Where Empires Collided*. For an account of the role Constantinople played as the center of the Ottoman Empire in its dealings with the world, see McNeill, "Ottoman Empire in World History"; Kasaba, *Ottoman Empire and the World Economy*. For a thorough examination of U.S.-Turkish cultural exchange, see, among others, Patrick, *Bosporus Adventure*.

67. Numerous illuminating studies have been conducted on the U.S.-Chinese relationship in the second half of the nineteenth century. Among them, certain works stand as classics. Written in the early 1980s, for example, Hunt's study *The Making of a Special Relationship* offers an insightful survey of U.S.-Chinese relations at the turn of the century, including the open-door policy. Following its trail, Hunter, *The Gospel of Gentility*, looks at

the U.S.-Chinese relationship from a cultural perspective, while McCormick, *China Market*, examines U.S.-Chinese relations in the sphere of commerce. McKee, *Chinese Exclusion vs. the Open Door Policy*, interrogates the Chinese Exclusion Act and the open-door policy as they were pursued by the United States during the late nineteenth and early twentieth centuries at the intersection of domestic and foreign policy. For a thorough understanding of the Chinese Exclusion Act in the wider context of alienation laws in the United States, see, among others, Lew-William, *Chinese Must Go*.

68. In "The Family That Plays Together," Braden asserts that following rapid urbanization and as the middle classes adopted a new, less rigid approach toward work ethic by century's end, entertainment gradually moved to the confines of the American home (145–46). In this vein, Melanie Dawson looks at home entertainment as a phenomenon characteristic of middle-class Americans during a wider time span, from 1850 to 1920. According to her, for many American families, home entertainment "formed a set of extravagant, bodily, and performative practices that were as much a cornerstone of the American cultural experience as piety, sentiment, and the intricacies of etiquette." Dawson, *Laboring to Play*, 1–2. Against this backdrop, and especially during the 1893 economic depression, puzzles had become exceedingly popular. The unsolved and unsolvable "15 Block Puzzle," for instance, was one of the most popular pastimes of the period. The inventor of its earlier version, Matthias Rice, reported that since the first box of the puzzle was sold in Boston, he barely managed to keep up with the orders. According to Jerry Slocum, the rage over this puzzle—later claimed by Sam Loyd as its original inventor—made its way to songs, plays, and political cartoons. On the controversy over the puzzle's origins, see Slocum, "Sam Loyd's Most Successful Hoax."

69. Wiegand discusses the factors that have direct and indirect influence on children's geographical memory of the world but, unlike the present study, does not talk about similar factors that might work on bigger or smaller scales, such as cities, rivers, mountains, oceans, and so on. Wiegand, *Places in the Primary School*, 75–76.

70. Foucault, *Order of Things*, xxii.

71. Kelen and Sundmark, *Nation in Children's Literature*, 2.

72. Kelen and Sundmark, *Nation in Children's Literature*, 2.

73. Needless to say, this utopian, predominantly white, and parochial vision of home as safe and supportive papers over domestic violence and trauma as dark and disruptive realities of the domestic sphere—that is, both American children's homes and the racially diverse nation.

Chapter Four

1. It is worth remembering that from April 30, 1895, until its last issue, the magazine was called *Harper's Round Table*, named after the Order of the Round Table, an American readers' club that appealed mainly to boys.

2. Over the course of the eighteenth century and even during the first half of the nineteenth, numerous imported and original manuals on the art of letter writing addressed Americans, from sailors and miners to gentlemen, young female secretaries, and schoolchildren. Late in the nineteenth century, letter-writing manuals addressing particular class, gender, and age groups entered the U.S. market. Volumes such as F. A. Brady's *New*

Letter-Writer, for the Use of Ladies addressed a primarily female adult audience. Among the manuals that addressed younger Americans, *The Youth's Letter-Writer, or The Epistolary Art Made Plain and Easy to Beginners*, by Mrs. John Farrar, provided samples on wide-ranging topics for a young audience.

3. M. M. Dodge, "Children's Magazines," 13–14.

4. According to Konstantin Dierks, as early as the eighteenth century, schoolteachers encouraged their students to correspond with each other as part of the school curriculum in order to ensure that letter writing established a place in their young lives. Dierks, *In My Power*, 240. By 1900, Dierks contends, British and American educators reacted to this trend and began to compile volumes that included letter-writing instructions exclusively addressed at secondary and even primary school children. According to Dierks, the first such volume compiled by an American was *Juvenile Letters: Being a Correspondence between Children from Eight to Fifteen Years of Age*, published in 1803. Authored by the Boston educator Caleb Bingham (1757–1817), this volume included a wide number of samples of child–adult and child–child correspondence. *Hints on Letter Writing, for the use of Academies and for Self-Instruction: Adapted from the French of the Author of "Golden Sands,"* another child-specific manual from the late nineteenth century, by Ella McMahon, focused on schoolchildren, especially young girls, and their correspondence with their mothers, sisters, and teachers.

5. For fine-grained studies on the material and nonmaterial characteristics and urges of the emerging middle class in the history of the United States, see, among others, Aron, *Ladies and Gentlemen of the Civil Service*; Barger and White, *Daguerreotype*; Aron, *Working at Play*; Blumin, *Emergence of the Middle Class*. For a multifaceted examination of middle-class American children in the nineteenth century, see, among others, O'Malley, *Making of the Modern Child*; Denisoff, *Nineteenth-Century Child*. Furthermore, for a thoroughgoing comparative study of the genesis and transformation of the middle classes from a transnational perspective, see López and Weinstein, *Making of the Middle Class*.

6. Mary Mapes Dodge is best remembered in her role as the editor of *St. Nicholas*, which was one of the most celebrated and long-lived American children's periodicals at the turn of the twentieth century. Before assuming the editorial responsibilities of *St. Nicholas*, Dodge had written a number of children's stories, the most famous of which, *Hans Brinker, or The Silver Skates* (1865), is believed to have been "the first American novel to take children inside a foreign culture." Rahn, "St. Nicholas and Its Friends: The Magazine-Child Relationship," in *St. Nicholas and Mary Mapes Dodge*, 107.

7. *Harper's Young People* 1, no. 10 (January 6, 1880).

8. As true as it is that adults could and did silence, alter, or at least redact children's voices through parenting and schooling, it is equally true that they played an active, empowering role in helping children's voices to be heard. In other words, adults played both silencing and amplifying roles. They sometimes functioned as ghostwriters, writing letters on behalf of their little sons and daughters or their younger siblings who wished to write a letter but were too young to write. For example, an unsigned letter from a young boy to *St. Nicholas* (September 1898) finished with the note, "My brother promised to write everything I said, but I can't read writing, and so I don't know if he has or if he has n't [*sic*]." This was also the case with the letter that had been "dictated by Jamie C." and printed in the November 1894 issue of *St. Nicholas*, and the letter by six-year-old Maude R. quoted in this chapter's epigraph. At times, adults also wrote notes on their children's letters to affirm the originality of

the ideas that the child had expressed or to ask the editors to mention a point about their behavior in their editorial note. Such was the case with Rachel Carson's story "A Battle in the Clouds," which won the silver medal for short stories in the St. Nicholas League. The submission included her story and a note in her mother's handwriting confirming that "this story was written without assistance, by my little ten-year-old daughter, Rachel." For a fine-grained study of Rachel Carson's career as an environmentalist and the origins of her career in fiction writing for *St. Nicholas* as a young girl, see her biography: Levine, *Rachel Carson*.

9. While these periodicals were more readily accessible to children who lived in the cities and who came from wealthier backgrounds, rural children and children of weaker economic backgrounds had access to them as well. Some children received the magazines through public libraries or as presents from their teachers or relatives, while others saved their pocket money and bought individual copies whenever they could. As Rahn asserts in the case of *St. Nicholas*, for example, there were "more ways than through subscriptions for children to become regular readers of *St. Nicholas*," including circulation of magazine issues among friends and relatives. Rahn, "St. Nicholas and Its Friends," 100–101.

10. Foucault, *"Society Must Be Defended,"* 255.

11. Here I deploy "foundational myth" in the sense that Heike Paul does. In Paul's understanding, myths are foundational beliefs "upon which constructions of the American nation have been based and which still determine contemporary discussions of US-American identities." Building on extensive existing scholarship on the subject, Paul characterizes these myths as those elements of U.S. national identity that have persisted throughout its history; that have provided Americans with cultural capital in the form of icons, rituals, and images; that have lent themselves to various historical moments and functioned as the context for various grand narratives about the "new world"; and that have made it at times reductively and discriminatorily possible for the white founders and inhabitants of this "new world" to imagine it as one exceptional, racially homogeneous nation. See the introduction to Paul, *The Myths That Made America*, 11–31. See also chapters 4 and 5 in the same volume, in which Paul discusses whiteness in relation to the U.S. presidency, in reference to the Founding Fathers' myth, and as part and parcel of the melting pot metaphor.

12. Racial invisibility starts at the phenotypical level of understanding "race" as an allegedly natural quality and a surface marker of this or that group of people, but it quickly transforms itself into a set of discriminatory social practices embedded in the ideology of racial hierarchy. Steve Garner maintains in his discussion of "racisms" that invisibility or racial color blindness should be viewed in relation to the following: whiteness as the norm, the power of whiteness to sustain itself as a norm, and the power of whiteness to mark the so-called nonwhite person individually as well as collectively invisible or negligible. Garner, *Racisms: An Introduction*, 118. To add to Garner's list, it should be noted that the disadvantages that whiteness subjects non-white Others to are more or less rendered invisible to secure the privileged position of whites in order to erase any sense of guilt and, further, to leave little room for dissent from others. In effect, effacing disadvantages in one direction also effaces the effects of privilege in the other, enabling white populations to believe that if whiteness as a privilege did not exist, then white privilege and nonwhite disadvantages would be nothing but a myth.

13. Of course, many immigrants who chose to live in the United States during the nineteenth century attempted to preserve their mother tongue as the spoken language at home and in their communities—attempts that more often than not failed in the face of the power

that public schooling exerted over their "un-American" lifestyles. In the case of Swedish immigrants, for instance, besides the more or less preserved national cuisine at home and in specialty food shops, what the Swedish community worried about, yet succeeded in retaining to some extent during the second half of the century, was their native language. The churches that Swedish immigrants belonged to were one social institution that proved instrumental in preserving Swedish as the social language of fraternity and belonging. However, as previously noted, the American public school stood against successful attempts by the church to foster Swedish as the everyday language within the Swedish community. For a detailed examination of the Swedish struggle against total Americanization of their lives, see Holmquist, *They Chose Minnesota*, 265–66.

14. The increase in the number of immigrants who entered the country during the 1890s had contributed to heightened nativist and anti-immigrant sentiments, reaching a boiling point across the nation during the mid-decade economic depression. For a thorough examination of U.S. immigration policies before, during, and after the 1890s, see Zolberg, *Nation by Design*. For fine-grained overviews of the literature on patriotism, nativist sentiments, and anti-immigration policies, see Higham, *Strangers in the Land*; Bennett, *Party of Fear*; Chan, *Entry Denied*; Bodnar, *Bonds of Affection*. For more recent studies on the subject, see Bean and Stevens, *America's Newcomers*; Behdad, *Forgetful Nation*; Portes and Rumbaut, *Immigrant America: A Portrait*; Ettinger, *Imaginary Lines*; Navarro, *Immigration Crisis*; Lee and Yung, *Angel Island*; Moloney, *National Insecurities*.

15. At the same time, only a slim percentage of children who might have spoken languages other than English at home were literate enough and had access to the magazines. While by the late nineteenth century a considerable number of immigrant families were sending their children to public schools, a juvenile periodical subscription—and even, at times, paying for children's education beyond the compulsory primary education—was seen as an unnecessary extravagance that had to be avoided.

16. It was, understandably, only in American children's letters from abroad or when they compared themselves or the nation to foreigners and foreign lands that Americanness was mentioned and emphasized.

17. Brubaker and Cooper, "Beyond 'Identity,'" 29.

18. It is likely that the late reverend mentioned in Guy P. B.'s letter was Rev. Edward Augustus Goodnough (1825–90), rector of Hobart Episcopal Church on the Oneida reservation for over thirty-five years. For a brief postmortem account of his work as a missionary among the Oneida, see "Obituary of Edward A. Goodnough."

19. Mintz, *Huck's Raft*, 171. According to Mintz, starting with the Carlisle Indian Industrial School established by Richard H. Pratt (1840–1924) in 1879, around 150 such boarding schools, mostly public and run by the Bureau of Indian Affairs, and some under contract with private Christian missions, existed around the country at the turn of the twentieth century. For firsthand accounts by Native American children who attended Carlisle Indian Industrial School, see Fear-Segal and Rose, *Carlisle Indian Industrial School*. For illuminating analyses of the history of native American boarding schools, see, among others, Churchill, *Kill the Indian, Save the Man*; Katanski, *Learning to Write "Indian"*; Trafzer, Keller, and Sisquoc, *Boarding School Blues*; Mauro, *Art of Americanization at the Carlisle Indian School*.

20. As early as 1609, Native American children were lawfully kidnapped by the agents of the Virginia Company of London to learn English and grow up Christian. For details on the

affairs of the Virginia Company of London, see Library of Congress, "Colonial Settlement, 1600s–1763." See also Price, *Love and Hate in Jamestown*.

21. In effect, to rename a person or a place is predicated on a far more violent (if only partly successful) act of first erasing their original name.

22. Passed into law by Congress on February 8, 1879, the Dawes Act granted the U.S. president the right to break up tribal land that was formerly communally owned by all tribe members and to allot it in smaller "sections" to those individual families and persons who were willing to start farming and renounce tribal communal rights. Complicating the terms under which U.S. citizenship was bestowed on willing Native Americans, Section 6 of the act stated that "every Indian born within the territorial limits of the United States to whom allotments shall have been made under the provisions of this act, or under any law or treaty, and every Indian born within the territorial limits of the United States who has voluntarily taken up, within said limits, his residence separate and apart from any tribe of Indians therein, and has adopted the habits of civilized life, is hereby declared to be a citizen of the United States, and is entitled to all the rights, privileges, and immunities of such citizens, whether said Indian has been or not, by birth or otherwise, a member of any tribe of Indians within the territorial limits of the United States without in any manner affecting the right of any such Indian to tribal or other property." To read the Dawes Act in its entirety, see "Transcript of Dawes Act (1887)."

23. For more on the subject of renaming, see chapter 3.

24. For nuanced examinations of the case within larger discussions on U.S. imperialism, see Burnett and Marshall, *Foreign in a Domestic Sense*; Kaplan, *Anarchy of Empire*.

25. Kaplan, *Anarchy of Empire*, 11, 12.

26. For an exhaustive account of the National Quarantine Act of 1893 with a focus on race, see Markel, *Quarantine!*

27. For a thorough description of how Staten Island functioned as a "waiting room" for immigrants to the United States, see Stephenson, "The Quarantine War."

28. Stephenson, "Quarantine War," 81.

29. During the course of the nineteenth century, the quarantine stations off New York Harbor witnessed several outbreaks of epidemics, which caused massive deaths among the immigrants. According to Ciarán Ó Murchadha, in 1847 alone, and in the middle of the Irish famine of 1845–52, between 16 and 20 percent of the Irish immigrants who entered the quarantine station died there. Murchadha, *Great Famine*, 152–53.

30. Another notable example is Douglas C. Baynton's brilliant discussion on the campaign against sign language in the United States. According to Baynton, the turn-of-the-century resistance against a unified sign language in the country—which coincided with heightened nativist sensations among the U.S. public—had roots in a peculiarly nativist belief popularized by Alexander Graham Bell and other prominent Americans. In their view, being deaf equaled being a foreigner, and therefore adopting a sign language in order for the deaf to have a means to communicate was comparable to an act of border crossing and of allowing languages other than English and people other than able-bodied white populations into the nation. Baynton, *Forbidden Signs*, 28–29. For a compelling examination of the issues of race and nonwhite migration in times of international epidemics, see McKiernan-González, *Fevered Measures*.

31. According to Stephenson, this attitude had resulted in heightened sentiments against the quarantine station, after which a frightened, angry local mob had set many of its buildings on fire on September 1 and 2, 1858. Stephenson, "Quarantine War," 79–81. See also the

report of the Executive Committee of Staten Island on the 1858 fire in *Facts and Documents Bearing upon the Legal and Moral Questions*.

32. Epple, Kaltmeier, and Lindner, *Entangled Histories*, 8. The authors also note that this approach is limited in its applicability to homogenize and convert comparisons into a universally applicable approach to historiography.

33. For in-depth discussions on national identity and the escalation of nativist feelings in response to foreignness at various moments in U.S. history, see Lee, *At America's Gates*, 19–22; Taylor Saito, "Model Minority, Yellow Peril"; Fender, *American and European National Identities*.

34. Love, "White Is the Color of Empire," 77.

35. For thoroughgoing accounts of the dense historical encounters between the United States and Hawaii before and after the annexation, see, among others, Coffman, *Island Edge of America*; Kauanui, *Hawaiian Blood*; Okihiro, *Island World*. For illuminating studies on the annexation of Hawaii within broader debates on the rise of U.S. empire, see, among others, Okihiro, *Pineapple Culture*; Goldstein, *Formations of United States Colonialism*; Paul Frymer, *Building an American Empire*.

36. Or, in the case of Sam Loyd's mechanical disappearance puzzles, to the mysterious void in between disks that rotated in opposite directions? For a detailed discussion of Loyd's disappearance puzzles, see chapter 2.

37. For an analysis of picnicking as the practice of going away from civilization and of Twain's understanding of it as the way Americans related to the world, see the introduction to the present volume.

38. Tuttle, *"Daddy's Gone to War,"* for example, looks at how American children experienced war during World War II. Furthermore, McEuen, *Making War, Making Women*, examines the gendered construction, marking, and policing of the home front on proxy battlegrounds during World War II and the role women and children played in defining the American home during the war.

39. Paul, *Mapping Migration*, 35.

40. Helen Nicholson, for instance, the twelve-year-old daughter of a mining engineer who lived with her family in Santa Elena, Mexico, told *St. Nicholas* (November 1897) that while picturesque and fond of music, "the Mexicans about here are very slow and lazy. When a Mexican is asked to do anything, the reply is always, 'Oh, *mañana*' (tomorrow)! This is why," she concluded dismissively, "Mexico is called 'The Land of Mañana.'"

41. For a detailed account of the battle, see Young, "Manila, the Philippines, Battle (1898)," in *War of 1898 and US Interventions*, edited by Benjamin R. Beede, 302–4.

42. Jameson, "Cognitive Mapping."

Conclusion

1. See Schulten, *Geographical Imagination in America*.
2. Ford, *Issues of War and Peace*, 165.
3. Twain, *Connecticut Yankee in King Arthur's Court*.
4. For an introduction to the notion of "geopolitical imaginaries" and its use in a historical context, see the special issue of *Forum for InterAmerican Research* on geopolitical imaginaries, especially Epple and Kramer, "Globalization, Imagination, Social Space."

5. According to Wiegand, multiple studies have examined Piaget's and Weil's 1951 study "The Development in Children of the Idea of the Homeland and of Relations with Other Countries" and have come to the conclusion that children have an inaccurate understanding of both the size and the shape of the earth. In "The Persian Columbus," it is the so-called Oriental emperor who displays similar childlike naivete. Wiegand, *Places in the Primary School*, 38–47. See also Jahoda, "Children's Concepts of Nationality"; Nussbaum and Novak, "Assessment of Children's Concepts of the Earth."

6. For discussions on the spatial unsettledness of the U.S. empire and the nation's attempts at centering world map projections on the United States, see chapter 1.

7. The notion of imperial knowledge as part and parcel of the institutionalization of imperial cultures and of imperial governance is further discussed, among others, in Barth and Cvetkovski, *Imperial Co-operation and Transfer*; Stoler, *Carnal Knowledge and Imperial Power*; Burton, *After the Imperial Turn*; Thompson, *Imperial Knowledge*; Richards, *Imperial Archive*.

8. Division of Insular Affairs, *Report of the United States Philippine Commission*.

9. Division of Insular Affairs, *Report of the United States Philippine Commission*, 283.

10. Atkinson, "Report of the General Superintendent of Public Instruction."

11. Atkinson, "Report of the General Superintendent of Public Instruction," 569. Note that while the geography exams included in the report did not have a regional focus, the history exams were divided into American History and General History.

12. Atkinson, "Report of the General Superintendent of Public Instruction," 571.

13. Atkinson, "Report of the General Superintendent of Public Instruction," 530.

14. Atkinson, "Report of the General Superintendent of Public Instruction," 561.

15. Atkinson, "Report of the General Superintendent of Public Instruction," 562.

16. Atkinson, "Report of the General Superintendent of Public Instruction," 560.

17. Atkinson, "Report of the General Superintendent of Public Instruction," 561.

18. Prescott F. Jernegan published several books specifically addressed at Filipino schoolchildren as well as teachers, including *A Short History of the Philippines, for Use in Philippine Schools* and the books he published with Philippine Education Publishers in Manila (see bibliography).

19. Jernegan, *Philippine Geography Primer*.

20. Jernegan, *Philippine Geography Primer*, iii.

21. Jernegan, *Philippine Geography Primer*, iv.

22. Jernegan, *Philippine Geography Primer*, vi.

23. Jernegan, *Philippine Geography Primer*, 33.

24. Jernegan, *Philippine Geography Primer*, 49.

25. Jernegan, *Philippine Geography Primer*, 51.

26. Jernegan, *Philippine Geography Primer*, 100–108.

27. Jernegan, *Philippine Geography Primer*, 109.

28. Twain, *Adventures of Huckleberry Finn*, 60. See also the accompanying illustration in the original volume titled "Taking a Rest," visualizing the scene where Huck "laid down in the bottom of the canoe and let her afloat . . . laid there and had a good rest and a smoke out of [his] pipe, looking away into the sky, not a cloud in it." Twain, *Adventures of Huckleberry Finn*, 59.

29. Zahra, "Imagined Noncommunities," 93.

30. Epple, "Globale Mikrogeschichte," 37–38; Foucault, *Power/Knowledge*, 55–58.

31. The term "Humanistic Geography" as discussed in Derek Gregory, et al., *The Dictionary of Human Geography*, 357.

32. Appadurai, *Modernity at Large*, 5.

33. For a fascinating approach to trips around the earth during and since the age of exploration, see Chaplin, *Round about the Earth*. Chaplin presents the early history of round-the-earth explorations through the dynamic interactions between sailors' bodies as microcosms navigating the world's open waters aboard wooden ships, considering ships as worlds in their own right.

34. Zahra deploys the term "national ambivalence" as a modification of the more common term "national indifference," which, she acknowledges, is not free from pejorative elements in terms—such as "backwardness," "regionalism," and "false consciousness"—that have been frequently deployed by the political elite to refer to the phenomenon. Zahra, "Imagined Noncommunities," 99.

35. Ultimately, the pivotal question that remains open and demands further research to be conducted in the even less systematically preserved archives of childhood in the margins of empires seems to be: How did children other than the privileged, urban, able-bodied, and white—that is, those who lived in the archipelagic expanse of the spatially unsettled U.S. empire; those who were rushed into adulthood due to a host of urgencies imposed on them by harsh living conditions; the children of those who had newly migrated to the United States; and African American, Asian American, and Native American children—engage with the geographical educations they received?

Bibliography

Primary Sources

BOOKS AND ARTICLES

Adams, Daniel. *Modern Geography, to which is Added a Brief Sketch of Ancient Geography; a Plain Method of Constructing Maps; and an Introduction to the Use of the Globes. Illustrated by Numerous Engravings. Accompanied by an Improved Atlas.* Boston: R.S. Davis, 1838.

Alphonse Island. "A Brief History of Alphonse Island and the Seychelles." Accessed November 2020, www.alphonse-island.com/en/blog/2017/07/brief-history-alphonse-island-seychelles.

———. "For the Ultimate Getaway." Accessed November 2020, www.alphonse-island.com/the-island.

"American Institute of Instruction, 1830 to 1906." *Journal of Education* 64, no. 1 (1906): 5–14.

Atkinson, Fred W. "Report of the General Superintendent of Public Instruction to the Secretary of Public Instruction for the Period from May 27, 1901, to October 1, 1901." In *Report of the United States Philippine Commission to the Secretary of War for the Period from December 1, 1900, to October 15, 1901*, 529–75. Washington, D.C.: Government Printing Office, 1901. https://archive.org/details/reportunitedsta02unkngoog.

Bingham, Caleb. *Juvenile Letters Being a Correspondence between Children from Eight to Fifteen Years of Age.* Boston: printed by David Carlisle for Caleb Bingham, 1803.

Borges, Jorge Luis, and Andrew Hurley. *Collected Fictions.* New York: Viking, 1998.

Bourne, Henry Eldridge. *The Teaching of History and Civics in the Elementary and the Secondary School.* New York: Longmans, Green, 1902. https://archive.org/stream/teachinghistory02bourgoog#page/n5/mode/2up.

Bowen, James A. *The Rand-McNally Primary School Geography.* Chicago: Rand McNally, 1894.

Brady, F. A. *New Letter-Writer, for the Use of Ladies: Embodying Letters on the Simplest Matters of Life, and on Various Subjects, with Applications for Situations, etc., and a Copious Appendix of Forms of Address, Bills, Receipts and Other Useful Matter.* New York: F.A. Brady, 1867.

Brigham, Albert Perry. "Physical Geography in Secondary Schools." *School Review* 5, no. 8 (1897).

Brooks, Arthur A. *Index to the Journal of Geography 1897 to 1921 (Including the Journal of School Geography, 1897–1901, and the Bulletin of the American Bureau of Geography, 1900–1901).* New York: National Council of Geography Teachers, 1922.

Clute, John J. *The School Geography.* New York: Samuel Wood & Sons, 1833.

Cole, Thomas. *A Pic-Nic Party.* 1846. Oil on canvas. Brooklyn Museum. www.brooklynmuseum.org/opencollection/objects/1356.

Commercial Relations of the United States with Foreign Countries. Washington, D.C.: Government Printing Office, 1872.
Confidential Price List and Telegraphy Code, McLoughlin Brothers. New York: McLoughlin Brothers, 1899.
Cornell, Sarah S. *Intermediate Geography: Forming Part Second of a Systematic Series of School Geographies.* New York: Appleton and Company, 1878.
Davis, William Morris. "Home Geography." *Journal of School Geography: A Monthly Journal Devoted to the Interests of the Common-School Teacher of Geography* 1 (1897): 2–7.
Dexter, Edwin Grant. *A History of Education in the United States.* New York: Macmillan, 1904.
Division of Insular Affairs. *Report of the United States Philippine Commission to the Secretary of War for the Period from December 1, 1900, to October 15, 1901.* Washington, D.C.: Government Printing Office, 1901. https://archive.org/details/reportunitedstao2unkngoog.
Dodge, Mary Mapes. "Children's Magazines." In *St. Nicholas and Mary Mapes Dodge: The Legacy of a Children's Magazine Editor, 1873–1905*, edited by Susan R. Gannon, Suzanne Rahn, and Ruth Anne Thompson, 13–17. Jefferson, N.C.: McFarland, 2004.
Dodge, Richard Elwood. *Home Geography and World Relations.* Chicago: Rand McNally, 1904.
———. *The Principles of Geography.* Chicago: Rand McNally, 1904.
Dodge, Richard Elwood, and Clara Barbara Kirchwey. *The Teaching of Geography in Elementary Schools.* Chicago: Rand McNally, 1913.
Dodge, Richard Elwood, and Earl Emmet Lackey. *Our Neighbors across the Seas.* New York: Rand McNally, 1932.
Du Bois, W. E. B. *Du Bois on Education.* Edited by Eugene F. Provenzo Jr. Lanham, Md.: Rowman and Littlefield, 2002.
Dwight, Nathaniel. *A Short but Comprehensive System of the Geography of the World: By Way of Question and Answer; Principally Designed for Children and Common Schools.* New York: Evert Duyckinck, 1808.
Facts and Documents Bearing upon the Legal and Moral Questions: Connected with the Recent Destruction of the Quarantine Buildings on Staten Island. New York: Wm. C. Bryant, 1858.
Farrar, John. *The Youth's Letter-Writer, or The Epistolary Art Made Plain and Easy to Beginners: Through the Example of Henry Moreton.* New York: R. Bartlett and S. Raynor, 1834.
Fear-Segal, Jacqueline, and Susan D. Rose, eds. *Carlisle Indian Industrial School: Indigenous Histories, Memories, and Reclamations.* Lincoln: University of Nebraska Press, 2018.
Goodrich, Samuel G. *Peter Parley's Method of Telling about Geography to Children with Nine Maps and Seventy-Five Engravings, Principally for the Use of Schools.* Burlington: C. Goodrich, 1830.
———. *Peter Parley's Method of Telling about the Geography of the Bible and Ancient Countries.* London: James S. Hodson, 1839.
———. *Peter Parley's Tales about America and Australia.* London: Darton and Clark, 1836.
———. *Peter Parley's Universal History on the Basis of Geography: For the Use of Families.* Boston: American Stationers, 1837.
———. *Tales about Europe, Asia, Africa, and America.* London: Thomas Tegg and Son, 1835.
———. *The Tales of Peter Parley about Africa.* Boston: Gray & Bowen and Carter & Hendee, 1830.

———. *The Tales of Peter Parley about America*. Boston: Samuel Goodrich, 1827.
Gordy, Wilbur F., and Willis I. Twitchell. *A Pathfinder in American History for the Use of Teachers, Normal Schools, and More Mature Pupils in Grammar Grades*. Boston: Lee and Shepard, 1893.
Hall, G. Stanley. *Adolescence: Its Psychology and Its Relations to Physiology, Anthropology, Sociology, Sex, Crime, Religion and Education*. New York: D. Appleton, 1904. https://archive.org/stream/adolescenceitspso1hall#page/n3/mode/2up.
———. "Child Study in Summer Schools." *Regents Bulletin* 28 (1894): 333–36.
———. *Educational Problems*. New York: D. Appleton, 1911.
Harper's School Geography, with Maps and Illustrations. New York: Harper and Brothers, 1886.
Hinsdale, B. A. "Questions in Geography and History." *Central School Journal* 15, no. 184 (1893): 8–9.
Hodgson Burnett, Frances. *The One I Knew Best of All: A Memory of the Mind of a Child*. New York: Charles Scribner's Sons, 1893.
Houston, Edwin J. *The Elements of Physical Geography: For the Use of Schools, Academies, and Colleges*. Philadelphia: Eldredge & Brother, 1892.
Hubbard, George D. "The Influence of the Presence, Discovery and Distribution of the Precious Metals in America on the Migration of People." *Bulletin of the American Geographical Society* 44, no. 2 (1912): 97–112.
———. "The Practice School Course in Geography in the Normal School." *Journal of Geography* 2, no. 8 (1903): 393–403.
Iddings, James. *The Monitor's Instructor, or A System of Practical Geography, of the United States of America, in Particular: also, Containing, I. Several Useful Geographical Definitions and Problems. II. The Ground Divisions of the Earth into Land and Water, Continents and Islands. III. The Local Situation of Empires, Kingdoms, &c. IV. The Rivers, Cities, Productions, Natural Curiosities, &c. of the United States, in Verse. V. Geographical Catechisms of the United States. For the British and Spanish Possessions in North America, and the West India Islands. VI. A Short Geographical Catechism for the Different Nations beyond the Sea.: The Whole Rendered Familiar for the Use of Schools*. Wilmington: William Black, 1804.
"Increasing Urbanization: Population Distribution by City Size, 1790 to 1890." United States Census Bureau. July 19, 2012. www.census.gov/dataviz/visualizations/005.
Jernegan, Prescott F. *1001 Questions and Answers on Philippine History and Civil Government*. Manila: Philippine Education, 1907.
———. *The Philippine Citizen: A Text-Book of Civics, Describing the Nature of Government, the Philippine Government, and the Rights and Duties of Citizens of the Philippines*. Manila: Philippine Education, 1907.
———. *Philippine Geography Primer*. Boston: D. C. Heath, 1906.
———. *A Short History of the Philippines, for Use in Philippine Schools*. New York: D. Appleton, 1905.
King, Charles F. *The Picturesque Geographical Reader*. Vol. 1, *At Home and at School*. Boston: Lee and Shepard, 1890.
Knox, Thomas Wallace. *The Boy Travellers in Australasia: Adventures of Two Youths in a Journey to the Sandwich, Marquesas, Society, Samoan and Feejee islands, and through the*

Colonies of New Zealand, New South Wales, Queensland, Victoria, Tasmania, and South Australia. New York: Harper & Brothers, 1888.

———. *The Boy Travellers in Central Europe: Adventures of Two Youths in a Journey through France, Switzerland, and Austria, with Excursions among the Alps of Switzerland and the Tyrol*. New York: Harper & Brothers, 1892.

———. *The Boy Travellers in the Far East*. New York: Harper & Brothers, 1879.

———. *The Boy Travellers in Great Britain and Ireland: Adventures of Two Youths in a Journey through Ireland, Scotland, Wales, and England, with Visits to the Hebrides and the Isle of Man*. New York: Harper & Brothers, 1891.

———. *The Boy Travellers in Mexico; Adventures of Two Youths in a Journey to Northern and Central Mexico, Campeachey, and Yucatan, with a Description of the Republics of Central America, and of the Nicaragua Canal*. New York: Harper & Brothers, 1889.

———. *The Boy Travellers in Northern Europe: Adventures of Two Youths in a Journey through Holland, Germany, Denmark, Norway, and Sweden, with Visits to Heligoland and the Land of the Midnight Sun*. New York: Harper & Brothers, 1892.

———. *The Boy Travellers in South America: Adventures of Two Youths in a Journey through Ecuador, Peru, Bolivia, Brazil, Paraguay, Argentine Republic, and Chili, with Descriptions of Patagonia and Tierra del Fuego, and Voyages upon the Amazon and La Plata Rivers*. New York: Harper & Brothers, 1886.

———. *The Boy Travellers in Southern Europe: Adventures of Two Youths in a Journey through Italy, southern France, and Spain, with Visits to Gibraltar and the Islands of Sicily and Malta*. New York: Harper & Brothers, 1893.

———. *The Boy Travellers in the Levant; Adventures of Two Youths in a Journey through Morocco, Algeria, Tunis, Greece, and Turkey, with Visits to the Islands of Rhodes and Cyprus, and the Site of Ancient Troy*. New York: Harper Brothers, 1894.

———. *The Boy Travellers in the Russian Empire: Adventures of Two Youths in a Journey in European and Asiatic Russia, with Accounts of a Tour across Siberia*. New York: Harper & Brothers, 1886.

Knox, Thomas Wallace, and Henry M. Stanley. *The Boy Travellers on the Congo: Adventures of Two Youths in a Journey with Henry M. Stanley "Through the Dark Continent."* New York: Harper & Brothers, 1887.

Library of Congress. "History of the Library of Congress." Accessed November 2020, www.loc.gov/about/history-of-the-library.

Loyd, Sam. *Sam Loyd's Cyclopedia of 5,000 Puzzles, Tricks, and Conundrums*. Pinnacle Books, 1914.

Mackenzie. James C. "The Report of the Committee of Ten." *School Review* 2 (1894): 146–55.

Mahan, Alfred T. *Influence of Sea Power upon History, 1660–1783*. Boston: Little, Brown, 1890.

The Making of a Great Magazine: Being an Inquiry into the Past and the Future of "Harper's Magazine." With Specimen Illustrations and a Partial Analysis of the Contents in Recent Years. New York: Harper and Brothers, 1889.

Malte-Brun, Conrad. *Précis de la géographie universelle: ou description de toutes les parties du monde, sur un plan nouveau, d'après les grandes divisions naturelles du globe; précédée de l'histoire de la géographie chez les peuples anciens et modernes, et d'une théorie générale de la géographie mathématique, physique et politique*. Paris: F. Buisson, 1810.

McLoughlin Bros.' Catalogue. New York: McLoughlin Bros., 1899.

"McLoughlin Bros. Collection." American Antiquarian Society. Accessed November 2020. www.americanantiquarian.org/mcloughlin-bros.

"McLoughlin Brothers Papers." De Grummond Children's Literature Collection, University of Southern Mississippi. Accessed November 2020, www.lib.usm.edu/legacy/degrum/public_html/html/research/findaids/DG0649f.html.

McMahon, Ella. *Hints on Letter Writing, for the Use of Academies and for Self-Instruction: Adapted from the French of the Author of "Golden Sands."* New York: Benziger Brothers, 1885.

Milton Bradley. *Catalogue of Games, Sectional Pictures, Toys, Puzzles, Blocks and Novelties, Made by Milton Bradley Company, 1889–1890*. New York: Wilson Brothers, 1889.

Mitchell, S. Augustus. *Mitchell's Geographical Reader: A System of Modern Geography, Comprising a Description of the World*. Philadelphia: Thomas, Cowperthwait, 1840.

Morse, Jedidiah. *The American Geography, or A View of the Present Situation of the United States of America ... Illustrated with Two Sheet Maps ... To which is Added, a Concise Abridgment of the Geography of the British, Spanish, French and Dutch Dominions in America, and the West Indies-of Europe, Asia and Africa*. Elizabethtown: Shepard Kollock, 1789.

———. *The American Universal Geography, or A View of the Present State of all the Empires, Kingdoms, States, and Republics in the Known World, and of the United States of America in Particular*. Boston: Isaiah Thomas and Ebenezer T. Andrews, 1792.

Morse, Sidney E. *A System of Geography for the Use of Schools: Illustrated with More than Fifty Cerographic Maps and Numerous Wood-Cut Engravings*. New York: Harper & Brothers, 1844.

Niles, Sanford. *The Complete Geography: Mathematical, Physical, Political*. Indianapolis: Indiana School Book, 1889.

"Obituary of Edward A. Goodnough, from the Wisconsin *Daily State Gazette*, Saturday Evening, 1 February 1890, p. 3." Accessed November 2020, www.public.coe.edu/~theller/soj/unc/tame-indians/goodnough-bio.html.

O'Sullivan, John L. "Annexation." *United States Magazine and Democratic Review* 17, no. 1 (1845): 5–10.

———. "The Great Nation of Futurity." *United States Magazine and Democratic Review* 6 (1839): 426–30.

"Peace Work at the American Institute of Instruction." *Advocate of Peace* 68, no. 8 (1906): 168–69.

Pratt, Mary Howe Smith, and Arnold Henry Guyot. *The Guyot Geographical Reader and Primer: A Series of Journeys Round the World*. New York: American Book, 1898.

Rand McNally & Company. *School and Library Globes*. Chicago: Rand McNally, ca. 1910s.

Ratzel, Friedrich. *Anthropo-Geographie, oder Grundzüge der Anwendung der Erdkunde auf die Geschichte*. Stuttgart: J. Engelhorn, 1882.

Roddy, H. Justin. *Complete Geography*. New York: American Book, 1902.

Roosevelt, Theodore. *The Rough Riders*. Mineola: Dover, 2006.

———. *The Strenuous Life: Essays and Addresses*. Mineola: Dover, 2009.

Stanley, Henry M. *Through the Dark Continent: Or, the Sources of the Nile, around the Great Lakes of Equatorial Africa and down the Livingstone River to the Atlantic Ocean*. New York: Harper and Brothers, 1878.

Stanley, Henry Morton, and Dorothy Stanley. *The Autobiography of Sir Henry Morton Stanley*. Cambridge: Cambridge University Press, 2012.

Stowe, Harriet Beecher. *First Geography for Children*. Boston: Phillips, Sampson, 1855.

Swinton, William. *Grammar-School Geography: Physical, Political, and Commercial*. New York: American Book, 1896.

Texas Institution for the Blind. *Annual Report*. Austin: Institution for the Blind, 1876.

"Transcript of Dawes Act (1887)." Our Documents Initiative. Accessed October 2020. www.ourdocuments.gov/doc.php?flash=false&doc=50&page=transcript.

Twain, Mark. *Adventures of Huckleberry Finn*. New York: Charles Webster, 1885.

———. *The Adventures of Tom Sawyer; Tom Sawyer Abroad; Tom Sawyer, Detective*. Edited by John C. Gerber, Paul Baender, and Terry Firkins. Berkeley: published for the Iowa Center for Textual Studies by the University of California Press, 1980.

———. *A Connecticut Yankee in King Arthur's Court*. Garden City, N.Y.: Nelson Doubleday, 1889.

———. *The Innocents Abroad, or The New Pilgrims' Progress: Being Some Account of the Steamship Quaker City's Pleasure Excursion to Europe and the Holy Land: With Descriptions of Countries, Nations, Incidents, and Adventures as They Appeared to the Author*. Hartford, Conn.: American, 1869.

———. *Tom Sawyer Abroad; Tom Sawyer Detective; and Other Stories*. New York: Harper and Brothers, 1896.

Washington, Booker T., Norman Barton Wood, and Fannie Barrier Williams. *A New Negro for a New Century; An Accurate and Up-to-Date Record of the Upward Struggles of the Negro Race. The Spanish-American War, Causes of It; Vivid Descriptions of Fierce Battles; Superb Heroism and Daring Deeds of the Negro Soldier . . . Education, Industrial Schools, Colleges, Universities and Their Relationship to the Race Problem*. Miami: Mnemosyne, 1969.

Webster, Noah. *A Grammatical Institute of the English Language: Comprising an Easy, Concise, and Systematic Method of Education, Designed for the Use of English Schools in America, in Three Parts, Part I Containing a New and Accurate Standard of Pronunciation*. Hartford, Conn.: printed by Hudson and Goodwin for the author, 1783.

———. *On the Education of Youth in America*. Boston: I. Thomas and E.T. Andrews, 1790.

"William McKinley: Second Inaugural Address (March 04, 1901)." In *Fellow Citizens: The Penguin Book of US Presidential Inaugural Addresses*, edited by Robert V. Remini and Terry Golway, 252–59. New York: Penguin Books, 2008.

Woodbridge, William C., and Emma Willard. *A System of Universal Geography: On the Principles of Comparison and Classification*. Hartford, Conn.: Oliver D. Cooke, 1824.

ONLINE COLLECTIONS

American Antiquarian Society Children's Books Collection
Bob Armstrong's Old Jigsaw Puzzles
British Library
David Rumsey Historical Map Collection
Division of Rare and Manuscript Collections, Cornell University Library
The George Washington University Museum
Internet Archive

Library Special Collections, Charles E. Young Research Library, UCLA
Lilly Library, Indiana University
The Newberry Library
Online collections of the New-York Historical Society
Project Gutenberg
PuzzleHistory.com Collections
Special Collections, Monroe C. Gutman Library, Harvard Graduate School of Education
The Strong National Museum of Play Online Collections

PERIODICALS

Harper's Round Table (monthly), 1895–1899
Harper's Young People (weekly), 1879–1893
Harper's Young People (monthly), 1893–1895
Journal of Geography, 1901–1908
Journal of School Geography, 1897–1901
St. Nicholas Illustrated (monthly), 1873–1943

Secondary Sources

Adams, Gretchen A. "'Pictures of the Vicious Ultimately Overcome by Misery and Shame': The Cultural Work of Early National Schoolbooks." In *Children and Youth in a New Nation*, edited by James Alan Marten, 149–69. New York: New York University Press, 2009.
Alasuutari, Maarit, Marleena Mustola, and Niina Rutanen. *Exploring Materiality in Childhood: Body, Relations and Space*. Abingdon: Routledge, 2021.
Alexander, Kristine. "Agency and Emotion Work." *Jeunesse: Young Peoples, Texts, Cultures* 7, no. 2 (2015): 120–28.
Alvarez, Joseph A. *Mark Twain's Geographical Imagination*. Newcastle upon Tyne: Cambridge Scholars, 2009.
Anderson, Benedict. "Census, Map, Museum." In *Imagined Communities: Reflections on the Origin and Spread of Nationalism*, 163–86. New York: Verso, 1991.
Appadurai, Arjun. *Modernity at Large: Cultural Dimensions of Globalization*. Minneapolis: University of Minnesota Press, 1996.
Ariès, Philippe. *Centuries of Childhood: A Social History of Family Life*. Translated by Robert Baldick. New York: Vintage, 1962.
Aron, Cindy S. *Ladies and Gentlemen of the Civil Service: Middle-Class Workers in Victorian America*. New York: Oxford University Press, 1987.
———. *Working at Play: A History of Vacations in the United States*. New York: Oxford University Press, 2001.
Ashcroft, Bill. *On Post-Colonial Futures: Transformations of a Colonial Culture*. London: Continuum, 2001.
Ashcroft, Bill, Gareth Griffiths, and Helen Tiffin. *Postcolonial Studies: The Key Concepts*. London: Routledge, 2013.
Bach, Rebecca Ann. *Colonial Transformations: The Cultural Production of the New Atlantic World, 1580–1640*. New York: Palgrave, 2000.

Bailey, Richard A. *Race and Redemption in Puritan New England*. New York: Oxford University Press, 2011.
Baker, Adam H. *Revival and Awakening: American Evangelical Missionaries in Iran and the Origins of Assyrian Nationalism*. Chicago: University of Chicago Press, 2015.
Baker, Anne. *Heartless Immensity: Literature, Culture, and Geography in Antebellum America*. Ann Arbor: University of Michigan Press, 2006.
Barger, M. Susan, and William Blaine White. *The Daguerreotype: Nineteenth-Century Technology and Modern Science*. Baltimore: Johns Hopkins University Press, 2000.
Barth, Volker, and Roland Cvetkovski. *Imperial Co-operation and Transfer, 1870–1930: Empires and Encounters*. London: Bloomsbury Academic, 2015.
Baudrillard, Jean. "Simulacra and Simulations." In *Jean Baudrillard, Selected Writings*, edited by Mark Poster, 166–84. Stanford, Calif.: Stanford University Press, 1988.
Baynton, Douglas C. *Forbidden Signs: American Culture and the Campaign against Sign Language*. Chicago: University of Chicago Press, 1996.
Bean, Frank, and Gillian Stevens. *America's Newcomers and the Dynamics of Diversity*. New York: Russell Sage, 2003.
Bederman, Gail. *Manliness and Civilization: A Cultural History of Gender and Race in the United States, 1880–1917*. Chicago: University of Chicago Press, 1995.
Behdad, Ali. *A Forgetful Nation: On Immigration and Cultural Identity in the United States*. Durham, N.C.: Duke University Press, 2005.
Bender, Thomas. *A Nation among Nations: America's Place in World History*. New York: Hill and Wang, 2006.
———. *Rethinking American History in a Global Age*. Berkeley: University of California Press, 2002.
Bennett, David Harry. *The Party of Fear: From Nativist Movements to the New Right in American History*. Chapel Hill: University of North Carolina Press, 1988.
Berger, Stefan, and Chris Lorenz. *The Contested Nation: Ethnicity, Class, Religion and Gender in National Histories*. Basingstoke, Palgrave Macmillan, 2008.
Bernstein, Robin. "Children's Books, Dolls, and the Performance of Race; or, the Possibility of Children's Literature." *PMLA* 126, no. 1 (2011): 160–69.
———. *Racial Innocence: Performing American Childhood from Slavery to Civil Rights*. New York: New York University Press, 2011.
Biddick, Kathleen. "The ABC of Ptolemy." In *Text and Territory: Geographical Imagination in the European Middle Ages*, edited by Sylvia Tomasch and Sealy Gilles, 268–94. Philadelphia: University of Pennsylvania Press, 1998.
Blumin, Stuart M. *The Emergence of the Middle Class: Social Experience in the American City, 1760–1900*. Cambridge: Cambridge University Press, 2002.
Bodnar, John E. *Bonds of Affection: Americans Define Their Patriotism*. Princeton, N.J.: Princeton University Press, 1996.
Bosse, David. "To Give a Strong and Pleasing Effect: Color on Early Maps and Prints." *Historic Deerfield* 13 (Summer 2012): 32–37.
Bowen, Margarita. *Empiricism and Geographical Thought: From Francis Bacon to Alexander von Humboldt*. Cambridge, Mass.: Cambridge University Press, 1981.
Bowersox, Jeff. *Raising Germans in the Age of Empire*. Oxford: Oxford University Press, 2013.

Braden, Donna R. "'The Family That Plays Together Stays Together': Family Pastimes and Indoor Amusements, 1890–1930." In *American Home Life, 1880–1930: A Social History of Spaces and Services*, edited by Jessica Foy and Thomas Schlereth, 145–61. Knoxville: University of Tennessee Press, 1994.

Brands, H. W. *The Reckless Decade: America in the 1890s*. Chicago: University of Chicago Press, 2002.

Brandt, Bettina. "Writing Political History after the 'Iconic Turn.'" In *Writing Political History Today*, edited by Willibald Steinmetz, Ingrid Gilcher-Holtey, and Heinz-Gerhard Haupt, 351–58. Frankfurt: Campus Verlag, 2013.

Bringuier, Jean Claude, and Jean Piaget. *Conversations with Jean Piaget*. Chicago: University of Chicago Press, 1980.

Brody, David. *Visualizing American Empire: Orientalism and Imperialism in the Philippines*. Chicago: University of Chicago Press, 2010.

Brubacher, John Seiler, and Willis Rudy. *Higher Education in Transition: A History of American Colleges and Universities, 1636–1968*. New York: Harper and Row, 1968.

Brubaker, Rogers, and Frederick Cooper. "Beyond 'Identity.'" *Theory and Society: Renewal and Critique in Social Theory* 29, no. 1 (2000): 1–47.

Brückner, Martin. *Early American Cartographies*. Chapel Hill: University of North Carolina Press, 2011.

———. *The Geographic Revolution in Early America: Maps, Literacy, and National Identity*. Chapel Hill: University of North Carolina Press, 2006.

———. "Lessons in Geography: Maps, Spellers, and Other Grammars of Nationalism in the Early Republic." *American Quarterly* 51, no. 2 (1999): 311–43.

———. "The Lithographed Map in Philadelphia: Innovation, Imitation, and Antebellum Consumer Culture." *Winterthur Portfolio* 48, no. 2/3 (2014): 139–62.

———. *The Social Life of Maps in America, 1750–1860*. Chapel Hill: University of North Carolina Press, 2017.

Brückner, Martin, and Hsuan L. Hsu, eds. *American Literary Geographies: Spatial Practice and Cultural Production, 1500–1900*. Newark: University of Delaware Press, 2007.

Burnett, Christina Duffy, and Burke Marshall. *Foreign in a Domestic Sense: Puerto Rico, American Expansion, and the Constitution*. Durham, N.C.: Duke University Press, 2001.

Burton, Antoinette. *After the Imperial Turn: Thinking with and through the Nation*. Durham, N.C.: Duke University Press, 2003.

Butler, Judith, and Elizabeth Weed, eds. *The Question of Gender: Joan W. Scott's Critical Feminism*. Bloomington: Indiana University Press, 2011.

Calloway, Colin G. *New Worlds for All: Indians, Europeans, and the Remaking of Early America*. Baltimore: Johns Hopkins University Press, 2013.

Capshaw Smith, Katharine, and Anna Mae Duane. *Who Writes for Black Children? African American Children's Literature before 1900*. Minneapolis: University of Minnesota Press, 2017.

Carpenter, Charles H. *History of American Schoolbooks*. Philadelphia: University of Pennsylvania Press, 1963.

Case, Jay Riley. *An Unpredictable Gospel: American Evangelicals and World Christianity, 1812–1920*. Oxford: Oxford University Press, 2012.

Center for InterAmerican Studies. "The Americas as Space of Entanglement (2013–2019)." Accessed July 2021, www.uni-bielefeld.de/einrichtungen/cias/forschung/entangled-americas.

Chan, Sucheng. *Entry Denied: Exclusion and the Chinese Community in America, 1882–1943*. Philadelphia: Temple University Press, 1991.

Chaplin, Joyce E. "Planetary Power? The United States and the History of Around-the-World Travel." *Journal of American Studies* 47, no. 1 (2013): 1–21.

———. *Round about the Earth: Circumnavigation from Magellan to Orbit*. New York: Simon and Schuster, 2012.

Chorley, Richard J., and Peter Haggett. *Frontiers in Geographical Teaching*. London: Methuen, 1970.

Chudacoff, Howard P. *Children at Play: An American History*. New York: New York University Press, 2007.

Churchill, Ward. *Kill the Indian, Save the Man: The Genocidal Impact of American Indian Residential Schools*. San Francisco: City Lights, 2005.

Clark, Beverly Lyon. *Kiddie Lit: The Cultural Construction of Children's Literature in America*. Baltimore: Johns Hopkins University Press, 2003.

Clark, Ian D., Luise Hercus, and Laura Kostanski. *Indigenous and Minority Place Names: Australian and International Perspectives*. Acton: Australian National University Press, 2014.

Coffman, Tom. *The Island Edge of America: A Political History of Hawai'i*. Honolulu: University of Hawaii Press, 2003.

Cohoon, Lorinda B. *Serialized Citizenships: Periodicals, Books, and American Boys, 1840–1911*. Lanham, Md.: Scarecrow Press, 2006.

Collins, Patricia Hill. "It's All in the Family: Intersections of Gender, Race, and Nation." In *Decentering the Center: Philosophy for a Multicultural, Postcolonial, and Feminist World*, edited by Uma Narayan and Sandra Harding, 156–76. Bloomington: Indiana University Press, 2000.

Costello, Matthew J. *The Greatest Puzzles of All Time*. New York: Prentice Hall Press, 1988.

Crain, Patricia. *Reading Children: Literacy, Property, and the Dilemmas of Childhood in Nineteenth-Century America*. Philadelphia: University of Pennsylvania Press, 2016.

———. *The Story of A: The Alphabetization of America from "The New England Primer" to "The Scarlet Letter."* Stanford, Calif.: Stanford University Press, 2000.

Cunningham, Hugh. *Children and Childhood in Western Society since 1500*. London: Routledge, 2017.

Daly, Reginald Aldworth. *Biographical Memoir of William Morris Davis, 1850–1934: Presented to the Academy at the Autumn Meeting, 1944*. Washington, D.C.: National Academy of Sciences, 1945.

Danesi, Marcel. *The Puzzle Instinct: The Meaning of Puzzles in Human Life*. Bloomington: Indiana University Press, 2002.

Darian-Smith, Kate, and Carla Pascoe. *Children, Childhood and Cultural Heritage*. London: Routledge, 2013.

Davis, Lance E., and Robert J. Cull. *Capital Movements, Markets, and Growth: 1820–1914*. Cambridge: Cambridge University Press, 2000.

Dawson, Melanie. *Laboring to Play: Home Entertainment and the Spectacle of Middle-Class Cultural Life, 1850–1920*. Tuscaloosa: University of Alabama Press, 2005.

Deleuze, Gilles, and Félix Guattari. *A Thousand Plateaus: Rhizomes*. London: A&C Black, 2004.
Denisoff, Dennis. *The Nineteenth-Century Child and Consumer Culture*. Aldershot, England: Ashgate, 2008.
D'haen, Theo, Paul Giles, Djelal Kadir, and Lois Parkinson Zamora, eds. *How Far Is America from Here? Selected Proceedings of the First World Congress of the International American Studies Association, 22–24 May 2003*. Amsterdam: Rodopi, 2005.
Dierks, Konstantin. *In My Power: Letter Writing and Communications in Early America*. Philadelphia: University of Pennsylvania Press, 2009.
Dobrin, Sidney I., and Kenneth B. Kidd. *Wild Things: Children's Culture and Ecocriticism*. Detroit: Wayne State University Press, 2004.
Dunbar, Gary S. *Geography: Discipline, Profession, and Subject Since 1870: An International Survey*. Dordrecht, Netherlands: Kluwer Academic, 2001.
Elleman, Bruce A. *International Competition in China, 1899–1991: The Rise, Fall, and Restoration of the Open Door Policy*. New York: Routledge, 2015.
Epple, Angelika. "Globale Mikrogeschichte. Auf dem Weg zu einer Geschichte der Relationen." In *Im Kleinen das Große suchen. Mikrogeschichte in Theorie und Praxis; Hans Haas zum 70. Geburtstag*, edited by Ewald Hiebl, Ernst Langthaler, and Hanns Haas, 37–47. Innsbruck: StudienVerlag, 2012.
———. "The Global, the Transnational, and the Subaltern: The Limits of History beyond the National Paradigm." In *Beyond Methodological Nationalism: Research Methodologies for Cross-Border Studies*, edited by Anna Amelina, Devrimsel D. Nergiz, Thomas Faist, and Nina Glick Schiller, 155–75. New York: Routledge, 2012.
Epple, Angelika, Olaf Kaltmeier, and Ulrike Lindner. *Entangled Histories: Reflecting on Concepts of Coloniality and Postcoloniality*. Leipzig: Leipziger Universitätsverlag, 2011.
Epple, Angelika, and Kirstin Kramer. "Globalization, Imagination, Social Space: The Making of Geopolitical Imaginaries." *Forum for InterAmerican Research* 9, no. 1 (2016).
Epple, Angelika, and Angelika Schaser, eds. *Gendering Historiography: Beyond National Canons*. Frankfurt am Main: Campus Verlag, 2009.
Ettinger, Patrick W. *Imaginary Lines: Border Enforcement and the Origins of Undocumented Immigration, 1882–1930*. Austin: University of Texas Press, 2009.
Eyal, Yonatan. *The Young America Movement and the Transformation of the Democratic Party, 1828–61*. Cambridge: Cambridge University Press, 2007.
Fass, Paula S. *The Routledge History of Childhood in the Western World*. London: Routledge, 2015.
Febvre, Lucien, Lionel Bataillon, Eleanor Gwen Mountford, and J. H. Paxton. *A Geographical Introduction to History*. London: Routledge, 2009.
Feinsod Cane, Aleta, and Susan Alves, eds. *The Only Efficient Instrument: American Women Writers and the Periodical, 1837–1916*. Iowa City: University of Iowa Press, 2001.
Fender, Stephen, ed. *American and European National Identities: Faces in the Mirror*. Staffordshire: Keele University Press, 1996.
Field, Hannah. *Playing with the Book: Victorian Movable Picture Books and the Child Reader*. Minneapolis: University of Minnesota Press, 2019.
Ford, Nancy Gentile. *Issues of War and Peace*. Westport, Conn: Greenwood Press, 2002.

Forman-Brunell, Miriam. *Made to Play House: Dolls and the Commercialization of American Girlhood, 1830–1930*. New Haven, Conn.: Yale University Press, 1993.
Foucault, Michel. *The Order of Things: An Archaeology of Human Sciences*. New York: Vintage Books, 1994.
———. *"Society Must Be Defended": Lectures at the Collège de France, 1975–1976*. Translated by David Macey. New York: Picador, 2003.
Foucault, Michel. *Power/Knowledge: Selected Interviews and Other Writings, 1972–1977*. Edited by Colin Gordon. New York: Pantheon Books, 1980.
Foy, Jessica H., and Thomas J. Schlereth. *American Home Life, 1880–1930: A Social History of Spaces and Services*. Knoxville: University of Tennessee Press, 1994.
Frymer, Paul. *Building an American Empire: The Era of Territorial and Political Expansion*. Princeton, N.J.: Princeton University Press, 2017.
Gallman, Robert E., and Stanley L. Engerman, eds. *The Cambridge Economic History of the United States*. Vol. 2, *The Long Nineteenth Century*. Cambridge: Cambridge University Press, 2000.
Gannon, Susan R. "'The Best Magazine for Children of All Ages': Cross-Editing St. Nicholas Magazine (1873–1905)." *Children's Literature* 25 (1997): 153–80.
Gannon, Susan R., Suzanne Rahn, and Ruth Anne Thompson. *St. Nicholas and Mary Mapes Dodge: The Legacy of a Children's Magazine Editor, 1873–1905*. Jefferson, N.C.: McFarland, 2004.
Gardner, Martin. *Mathematical Puzzles of Sam Loyd*. New York: Dover, 1959.
———. *More Mathematical Puzzles of Sam Loyd*. New York: Dover, 1960.
Garner, Steve. *Racisms: An Introduction*. Los Angeles: Sage, 2010.
Gerzina, Gretchen. *Frances Hodgson Burnett: The Unexpected Life of the Author of "The Secret Garden."* New Brunswick, N.J.: Rutgers University Press, 2004.
Gillard, D. R. "Salisbury's African Policy and the Heligoland Offer of 1890." *English Historical Review* 75, no. 297 (1960): 631–53.
Ginzburg, Carlo. "Clues: Roots of an Evidential Paradigm." In *Clues, Myths, and the Historical Method*, translated by John Tedeschi and Anne C. Tedeschi, 96–125. Baltimore: Johns Hopkins University Press, 2013.
Gleason, Mona. "Avoiding the Agency Trap: Caveats for Historians of Children, Youth and Education." *History of Education* 45, no. 4 (2016): 446–59.
Gocker, Ben. "Loyd's Puzzles." March 13, 2010, *Brooklynology* (blog). Brooklyn Public Library. www.bklynlibrary.org/blog/2010/03/13/loyds-puzzles.
Godley, Andrew C. "Foreign Investment." In *Encyclopedia of the United States in the Nineteenth Century*, edited by Paul Finkelman. New York: Charles Scribner's Sons, 2001.
Goga, Nina, Lykke Guanio-Uluru, Bjørg Oddrun Hallås, and Aslaug Nyrnes. *Ecocritical Perspectives on Children's Texts and Cultures: Nordic Dialogues*. Cham: Palgrave Macmillan, 2018.
Goldstein, Alyosha. *Formations of United States Colonialism*. Durham, N.C.: Duke University Press, 2014.
Gould, Peter. *Mental Maps*. Harmondsworth: Penguin Books, 1974.
———. "On Mental Maps." In *Image and Environment: Cognitive Mapping and Spatial Behavior*, edited by Roger M. Downs and David Stea, 182–220. New Brunswick, N.J.: Aldine Transaction, 2017.

Grabham, Emily. *Intersectionality and Beyond: Law, Power and the Politics of Location.* Abingdon: Routledge-Cavendish, 2009.

Graves, Joseph L. *The Emperor's New Clothes: Biological Theories of Race at the Millennium.* New Brunswick, N.J.: Rutgers University Press, 2001.

Greenberg, Amy S. *Manifest Manhood and the Antebellum American Empire.* Cambridge: Cambridge University Press, 2005.

Greenberg, Kenneth S. *Nat Turner: A Slave Rebellion in History and Memory.* Oxford: Oxford University Press, 2003.

Grieshaber, Susan. *Rethinking Parent and Child Conflict.* New York: Routledge, 2004.

Hackett, Abigail, Lisa Procter, and Julie Seymour. *Children's Spatialities: Embodiment, Emotion and Agency.* Houndmills: Palgrave Macmillan, 2015.

Haggett, Peter. *Geography: A Global Synthesis.* Harlow: Pearson Hall, 2001.

Haggis, Jane. "Gendering Colonialism or Colonising Gender? Recent Women's Studies Approaches to White Women and the History of British Colonialism." *Women's Studies International Forum* 13, no. 1/2 (1990): 105–15.

Hale, Jonathan. "Cognitive Mapping: New York vs. Philadelphia." In *The Hieroglyphics of Space: Reading and Experiencing Modern Metropolis,* edited by Neil Leach, 31–42. London: Routledge, 2002.

Hall, Roger A. *Performing the American Frontier, 1870–1906.* Cambridge: Cambridge University Press, 2001.

Hamilton, Richard F. *America's New Empire: The 1890s and Beyond.* London: Routledge, 2010.

Hannas, Linda. *The Jigsaw Book.* New York: Dial Press, 1981.

Harris, Victoria. "Sex on the Margins: New Directions in the Historiography of Sexuality and Gender." *Historical Journal* 53, no. 4 (2010): 1085–1104.

Harvey, David. "Between Space and Time: Reflections on the Geographical Imagination." *Annals of the Association of American Geographers* 80, no. 3 (1990): 418–34.

———. *The Condition of Postmodernity: An Inquiry into the Origins of Cultural Change.* Oxford: Wiley-Blackwell, 1989.

———. *The Limits to Capital.* Oxford: Blackwell, 1982.

Haschemi Yekani, Elahe. *The Privilege of Crisis: Narratives of Masculinities in Colonial and Postcolonial Literature, Photography, and Film.* Frankurt am Main: Campus Verlag, 2011.

Hawes, Joseph M., and N. Ray Hiner. "Reflections on the History of Children and Childhood in the Postmodern Era." In *Major Problems in the History of American Families and Children,* edited by Anya Jabour, 23–29. Boston: Houghton Mifflin, 2005.

Hawkins, Stephanie L. *American Iconographic: National Geographic, Global Culture, and the Visual Imagination.* Charlottesville: University of Virginia Press, 2010.

Healy, David. *US Expansionism: The Imperialist Urge in the 1890s.* Madison: University of Wisconsin Press, 2011.

Hewes, Lauren B., Justin G. Schiller, Laura E. Wasowicz, and Layla Haveles Hopper. *Radiant with Color and Art: McLoughlin Brothers and the Business of Picture Books, 1858–1920.* New Castle, Del.: Oak Knoll, 2017.

Heywood, Colin. *A History of Childhood: Children and Childhood in the West from Medieval to Modern Times.* Cambridge: Polity, 2018.

Higham, John. *Strangers in the Land: Patterns of American Nativism, 1860–1925.* New Brunswick, N.J.: Rutgers University Press, 1955.

Hodgkiss, Alan G. *Discovering Antique Maps*. Buckinghamshire: Shire, 2007.
Hoganson, Kristin L. *Fighting for American Manhood: How Gender Politics Provoked the Spanish-American and Philippine-American Wars*. New Haven, Conn.: Yale University Press, 2000.
Holborn, Hajo. *A History of Modern Germany: 1840–1945*. Princeton, N.J.: Princeton University Press, 1982.
Holmquist, June Drenning. *They Chose Minnesota: A Survey of the State's Ethnic Groups*. St. Paul: Minnesota Historical Society Press, 2004.
Holt-Jensen, Arild. *Geography: History and Concepts*. Los Angeles: Sage, 2009.
Honeck, Mischa. *Our Frontier Is the World: The Boy Scouts in the Age of American Ascendancy*. Ithaca, N.Y.: Cornell University Press, 2018.
Hooper, Charlotte. *Manly States: Masculinities, International Relations, and Gender Politics*. New York: Columbia University Press, 2001.
Hoose, Phillip. *We Were There, Too! Young People in U.S. History*. New York: Melanie Kroupa Books, 2001.
Hsu, Hsuan L. "Body Counts and Comparative Imperialism." In *Sitting in Darkness: Mark Twain's Asia and Comparative Racialization*, 139–66. New York: New York University Press, 2015.
———. *Geography and the Production of Space in Nineteenth-Century American Literature*. Cambridge: Cambridge University Press, 2010.
Huff, Anne Sigismund. *Mapping Strategic Thought*. Chichester: Wiley, 1990.
Hunt, Michael H. *The Making of a Special Relationship: The United States and China to 1914*. New York: Columbia University Press, 1983.
Hunter, Jane. *The Gospel of Gentility: American Women Missionaries in Turn-of-the-Century China*. New Haven, Conn.: Yale University Press, 1984.
Jacob, Christian. *The Sovereign Map: Theoretical Approaches in Cartography throughout History*, translated by Tom Conley and Edward H. Dahl. Chicago: University of Chicago Press, 2006.
Jahoda, Gustav. "Children's Concepts of Nationality: A Critical Study of Piaget's Stages." *Child Development* 35, no. 4 (1964): 1081–92.
James, Preston E., and Clarence Fielden Jones. *American Geography: Inventory and Prospect*. Syracuse: Syracuse University Press, 1954.
Jameson, Fredric. "Cognitive Mapping." In *Marxism and the Interpretation of Culture*, edited by Cary Nelson and Lawrence Grossberg, 347–60. Urbana: University of Illinois Press, 1988.
———. *Postmodernism, or The Cultural Logic of Late Capitalism*. Durham, N.C.: Duke University Press, 1991.
Jeal, Tim. *Stanley: The Impossible Life of Africa's Greatest Explorer*. New Haven, Conn.: Yale University Press, 2007.
Jehlen, Myra. *American Incarnation: The Individual, the Nation, and the Continent*. Cambridge, Mass.: Harvard University Press, 1986.
Johnson, Clifton. *Old-Time Schools and School-Books: With a New Introduction by Carl Withers*. New York: Dover, 1963.
Jordan, Benjamin René. *Modern Manhood and the Boy Scouts of America*. Chapel Hill: University of North Carolina Press, 2016.

Jordanova, Ludmilla. "Gender and the Historiography of Science." *British Journal for the History of Science* 26 (1993): 469–83.

Kaplan, Amy. *The Anarchy of Empire in the Making of US Culture*. Cambridge, Mass: Harvard University Press, 2002.

———. "Manifest Domesticity." *American Literature* 70, no. 3 (1998): 581–606.

Kasaba, Resat. *The Ottoman Empire and the World Economy: The Nineteenth Century*. Albany: SUNY Press, 1988.

Katanski, Amelia V. *Learning to Write "Indian": The Boarding-School Experience and American Indian Literature*. Norman: University of Oklahoma Press, 2007.

Kauanui, J. Kēhaulani. *Hawaiian Blood: Colonialism and the Politics of Sovereignty and Indigeneity*. Durham, N.C.: Duke University Press, 2008.

Keighren, Innes M. *Bringing Geography to Book: Ellen Semple and the Reception of Geographical Knowledge*. London: I. B. Tauris, 2010.

Kelen, Christopher, and Björn Sundmark. *The Nation in Children's Literature: Nations of Childhood*. New York: Routledge, 2013.

Kelly, R. Gordon. *Children's Periodicals of the United States*. Westport: Greenwood Press, 1984.

Kimmel, Michael S. *Manhood in America: A Cultural History*. New York: Free Press, 1996.

King, Desmond S. *The Liberty of Strangers: Making the American Nation*. New York: Oxford University Press, 2004.

Kliebard, Herbert M. *The Struggle for the American Curriculum, 1893–1958*. New York: Routledge and Kegan Paul, 1986.

Koelsch, William A. "Academic Geography, American Style: An Institutional Perspective." In *Geography: Discipline, Profession, and Subject*, edited by Gary S. Dunbar, 245–79. Dordrecht, Netherlands: Kluwer Academic Publishers, 2001.

LaFeber, Walter. *The New Empire: An Interpretation of American Expansion, 1860–1898*. Ithaca, N.Y.: Cornell University Press, 1963.

LaFeber, Walter, and Warren I. Cohen. *The New Cambridge History of American Foreign Relations: The American Search for Opportunity, 1865–1913*. Cambridge: Cambridge University Press, 2013.

Lee, Erika. *At America's Gates: Chinese Immigration during the Exclusion Era, 1882–1943*. Chapel Hill: University of North Carolina Press, 2007.

Lee, Erika, and Judy Yung. *Angel Island: Immigrant Gateway to America*. Oxford: Oxford University Press, 2010.

Levine, Ellen S. *Rachel Carson: A Twentieth-Century Life*. New York: Viking, 2008.

Lew-Williams, Beth. *The Chinese Must Go: Violence, Exclusion, and the Making of the Alien in America*. Cambridge, Mass.: Harvard University Press, 2018.

Library of Congress. "Colonial Settlement, 1600s–1763." Library of Congress. Accessed November 2020. www.loc.gov/classroom-materials/united-states-history-primary-source-timeline/colonial-settlement-1600-1763.

Little, Douglas. *American Orientalism: The United States and the Middle East since 1945*. Chapel Hill: University of North Carolina Press, 2002.

Littlefield, Henry M. "The Wizard of Oz: Parable on Populism." *American Quarterly* 16, no. 1 (1964): 47–58.

Livingston, David N. *The Geographical Tradition: Episodes in the History of a Contested Enterprise*. Oxford: Blackwell, 1992.

López, A. Ricardo, and Barbara Weinstein. *The Making of the Middle Class: Toward a Transnational History of the Middle Class.* Durham, N.C.: Duke University Press, 2012.

Love, Eric. "White Is the Color of Empire: The Annexation of Hawaii in 1898." In *Race, Nation, and Empire in American History*, edited by James T. Campbell, Matthew Pratt Guterl, and Robert G. Lee, 75–102. Chapel Hill: University of North Carolina Press, 2007.

Lutz, Catherine, and Jane Lou Collins. *Reading National Geographic.* Chicago: University of Chicago Press, 1993.

Lynch, Kevin. *The Image of the City.* Cambridge: MIT Press, 1960.

MacCabe, Colin. Preface to *The Geopolitical Aesthetic: Cinema and Space in the World System*, by Frederic Jameson, ix–xvi. Bloomington: Indiana University Press, 1995.

Mackey, Margaret. *One Child Reading: My Auto-Bibliography.* Edmonton: University of Alberta Press, 2016.

Malpas, J. E. *Heidegger and the Thinking of Place: Explorations in the Topology of Being.* Cambridge: MIT Press, 2012.

Markel, Howard. *Quarantine! East European Jewish Immigrants and the New York City Epidemics of 1892.* Baltimore: Johns Hopkins University Press, 1999.

Marsh, Jackie, and Elaine Millard. *Literacy and Popular Culture: Using Children's Culture in the Classroom.* London: Paul Chapman, 2000.

Marx, Leo. "The Pilot and the Passenger: Landscape Conventions and the Style of *Huckleberry Finn*." In *On Mark Twain*, edited by Louis J. Budd and Edwin Harrison Cady, 53–70. Durham, N.C.: Duke University Press, 1987.

Mauro, Hayes Peter. *The Art of Americanization at the Carlisle Indian School.* Albuquerque: University of New Mexico Press, 2011.

Mayar, Mahshid. "'Playes Print the Letter': Childhood, Temporality, and the Historical Archive." *Journal of the History of Childhood and Youth* 14, no. 1 (2021).

———. "Verbs of Violence: 19th-Century Jigsaw Puzzles, Otherness, and American Childhood." Society for the History of Children and Youth. November 6, 2019. www.shcy.org/features/commentaries/verbs-of-violence-19th-century-jigsaw-puzzles-otherness-and-american-childhood.

———. "What on Earth! Slated Globes, School Geography and Imperial Pedagogy." *European Journal of American Studies* 15, no. 2 (2020): 1–19.

McCormick, Thomas J. *China Market: America's Quest for Informal Empire, 1893–1901.* Chicago: Quadrangle Books, 1967.

McCoy, Alfred W., and Francisco A. Scarano, eds. *Colonial Crucible: Empire in the Making of the Modern American State.* Madison: University of Wisconsin Press, 2009.

McEuen, Melissa A. *Making War, Making Women: Femininity and Duty on the American Home Front, 1941–1945.* Athens: University of Georgia Press, 2011.

McGrath, Ann. "Playing Colonial: Cowgirls, Cowboys, and Indians in Australia and North America." *Journal of Colonialism and Colonial History* 2, no. 1 (2001).

McKee, Delber L. *Chinese Exclusion vs. the Open Door Policy, 1900–1906: Clashes over China Policy in the Roosevelt Era.* Detroit: Wayne State University, 1977.

McKenzie, Andrea. "A 'Revolutionary' War?" In *Children's Literature and Culture of the First World War*, edited by Lissa Paul, Rosemary R. Johnston, and Emma Short, 60–76. New York: Routledge, 2016.

McKiernan-González, John. *Fevered Measures: Public Health and Race at the Texas-Mexico Border, 1848–1942.* Durham, N.C.: Duke University Press, 2012.
McNeill, William H. "The Ottoman Empire in World History." In *The Ottoman State and Its Place in World History,* edited by Kemal H. Karpat, 34–46. Leiden: E. J. Brill, 1974.
McNenly, Linda Scarangella. *Native Performers in Wild West Shows: From Buffalo Bill to Euro Disney.* Norman: University of Oklahoma Press, 2012.
Melton, Jeffrey Alan. *Mark Twain, Travel Books, and Tourism: The Tide of a Great Popular Movement.* Tuscaloosa: University of Alabama Press, 2002.
Miller, Susan A. "Assent as Agency in the Early Years of the Children of the American Revolution." *Journal of the History of Childhood and Youth* 9, no. 1 (2016): 48–65.
Mintz, Steven. *Huck's Raft: A History of American Childhood.* Cambridge, Mass.: Belknap Press of Harvard University Press, 2004.
Moloney, Deirdre M. *National Insecurities: Immigrants and U.S. Deportation Policy since 1882.* Chapel Hill: University of North Carolina Press, 2012.
Monmonier, Mark. *How to Lie with Maps.* Chicago: University of Chicago Press, 2018.
———. *Rhumb Lines and Map Wars: A Social History of the Mercator Projection.* Chicago: University of Chicago Press, 2004.
Moore, Gregory. *Defining and Defending the Open Door Policy: Theodore Roosevelt and China, 1901–1909.* Lanham, Md.: Lexington Books, 2015.
Moruzi, Kristine, Nell Musgrove, and Carla Pascoe Leahy. "Hearing Children's Voices: Conceptual and Methodological Challenges." In *Children's Voices from the Past: New Historical and Interdisciplinary Perspectives,* edited by Kristine Moruzi, Nell Musgrove, and Carla Pascoe Leahy, 1–25. Cham: Palgrave McMillan, 2019.
Moses, Lester G. *Wild West Shows and the Images of American Indians: 1883–1933.* Albuquerque: University of New Mexico Press, 1996.
Mott, Frank Luther. *A History of American Magazines: 1885–1905.* Cambridge, Mass.: Harvard University Press, 1957.
Murphy, Alexander B. "Geography's Place in Higher Education in the United States." *Journal of Geography in Higher Education* 31, no. 1 (2007): 121–41.
Navarro, Armando. *The Immigration Crisis: Nativism, Armed Vigilantism, and the Rise of a Countervailing Movement.* Lanham, Md.: AltaMira Press, 2009.
Nelson, Cary, and Lawrence Grossberg. *Marxism and the Interpretation of Culture.* Urbana: University of Illinois Press, 1988.
Nesmith, Chris. "Young Americans Abroad: Jacob Abbott's Rollo on the Grand Tour and Nineteenth-Century Travel Series Books." In *Internationalism in Children's Series,* edited by Karen Sands-O'Connor and Marietta A. Frank, 19–37. London: Palgrave Macmillan, 2014.
Nodelman, Perry. "The Other: Orientalism, Colonialism, and Children's Literature." *Children's Literature Association Quarterly* 17, no. 1 (1992): 29–35.
Norcia, Megan A. *Gaming Empire in Children's British Board Games, 1836–1860.* London: Routledge, 2019.
———. "Playing Empire: Children's Parlor Games, Home Theatricals, and Improvisational Play." *Children's Literature Association Quarterly* 29, no. 4 (2004): 294–314.
———. "Puzzling Empire: Early Puzzles and Dissected Maps as Imperial Heuristics." *Children's Literature* 37, no. 1 (2009): 1–32.

———. *X Marks the Spot: Women Writers Map the Empire for British Children, 1790–1895*. Athens: Ohio University Press, 2010.

Ntamushobora, Faustin. *Education for Holistic Transformation in Africa*. Eugene: Wipf and Stock, 2015.

Nussbaum, Joseph, and Joseph D. Novak. "An Assessment of Children's Concepts of the Earth Utilising Structured Interviews." *Science Education* 60, no. 4 (1976): 535–50.

Okihiro, Gary Y. *Island World: A History of Hawai'i and the United States*. Berkeley: University of California Press, 2008.

———. *Pineapple Culture: A History of the Tropical and Temperate Zones*. Berkeley: University of California Press, 2010.

O'Malley, Andrew. *The Making of the Modern Child: Children's Literature and Childhood in the Late Eighteenth Century*. New York: Routledge, 2003.

Ó Murchadha, Ciarán. *The Great Famine: Ireland's Agony, 1845–1852*. London: Bloomsbury, 2011.

Onion, Rebecca. *Innocent Experiments: Childhood and the Culture of Popular Science in the United States*. Chapel Hill: University of North Carolina Press, 2016.

Parker, Michael. *John Winthrop: Founding the City upon a Hill*. New York: Routledge, 2014, 191–93.

Patrick, Mary Mills. *A Bosporus Adventure: Istanbul (Constantinople) Woman's College, 1871–1924*. Stanford, Calif.: Stanford University Press, 1934.

Paul, Heike. *Mapping Migration: Women's Writing and the American Immigrant Experience from the 1950s to the 1990s*. Heidelberg: C. Winter, 1999.

———. *The Myths That Made America: An Introduction to American Studies*. Bielefeld: Transcript, 2014.

Peltonen, Matti. "What Is Micro in Microhistory?" In *Theoretical Discussions of Biography: Approaches from History, Microhistory, and Life Writing*, edited by Hans Renders and Binne de Haan, 105–18. Leiden: Brill, 2014.

Phillips, Christopher. *Freedom's Port: The African American Community of Baltimore, 1790–1860*. Urbana: University of Illinois Press, 1997.

Phillips, Richard. *Mapping Men and Empire: A Geography of Adventure*. London: Routledge, 1997.

Piaget, Jean. *The Child's Conception of the World*. Translated by Joan Tomlinson and Andrew Tomlinson. London: Kegan Paul, Trench, Trubner, 1929. https://archive.org/details/childsconceptiono1piag.

Piaget, Jean, and Bärbel Inhelder. *The Child's Conception of Space*. Translated by F. J. Langdon and J. L. Lunzer. New York: W. W. Norton, 1967.

Piaget, Jean, and Anne-Marie Weil. "The Development in Children of the Idea of the Homeland and of Relations with Other Countries." *International Social Science Bulletin* 3 (1951): 561–78. http://unesdoc.unesco.org/images/0005/000593/059379eo.pdf#59388.

Pickard, Sid, and Sam Loyd. *The Puzzle King: Sam Loyd's Chess Problems and Selected Mathematical Puzzles*. Dallas: Pickard & Sons, 1996.

Portes, Alejandro, and Rubén G. Rumbaut. *Immigrant America: A Portrait*. Berkeley: University of California Press, 2006.

Pratt, Mary Louise. "Science, Planetary Consciousness, Interiors." In *Imperial Eyes: Travel Writing and Transculturation*. 15–37. London: Routledge, 1992.

Price, David A. *Love and Hate in Jamestown: John Smith, Pocahontas, and the Start of a New Nation*. New York: Vintage Books, 2005.

Quirk, Tom, and Gary Scharnhorst, eds. *American History through Literature, 1870–1920*. Detroit: Charles Scribner's Sons/Thomson Gale, 2006.

Rankin, William. *After the Map: Cartography, Navigation, and the Transformation of Territory in the Twentieth Century*. Chicago: University of Chicago Press, 2016.

Ratzel, Friedrich. *The Sea as a Source of the Greatness of a People: A Political-Geographical Study*. Munich: R. Oldenbourg, 1900.

Redcay, Anna M. "'Live to Learn and Learn to Live': The St. Nicholas League and the Vocation of Childhood." *Children's Literature* 39, no. 1 (2011): 58–84.

Reddin, Paul. *Wild West Shows*. Urbana: University of Illinois Press, 1999.

Richards, Thomas. *The Imperial Archive: Knowledge and the Fantasy of Empire*. London: Verso, 1993.

Roberts, Brian Russell, and Michelle Ann Stephens. "Archipelagic American Studies: Decontinentalizing the Study of American Culture." In *Archipelagic American Studies*, edited by Brian Russell Roberts and Michelle Ann Stephens, 1–54. Durham, N.C.: Duke University Press, 2017.

Rollo, Toby. "Feral Children: Settler Colonialism, Progress, and the Figure of the Child." *Settler Colonial Studies* 8, no. 1 (2018): 60–79.

Rothenberg, Tamar Y. *Presenting America's World: Strategies of Innocence in National Geographic Magazine, 1888–1945*. Aldershot: Ashgate, 2007.

Said, Edward W. *Culture and Imperialism*. New York: Knopf, 1993.

———. *Orientalism*. New York: Vintage Books, 1978.

Samuels, Peggy, and Harold Samuels. *Teddy Roosevelt at San Juan: The Making of a President*. College Station: Texas A&M University Press, 1997.

Samuels, Shirley C. *Romances of the Republic: Women, the Family, and Violence in the Literature of the Early American Nation*. New York: Oxford University Press, 1996.

Sánchez-Eppler, Karen. *Dependent States: The Child's Part in Nineteenth-Century American Culture*. Chicago: University of Chicago Press, 2005.

———. "In the Archives of Childhood." In *The Children's Table: Childhood Studies and the Humanities*, edited by Anna Mae Duane, 213–37. Athens: University of Georgia Press, 2013.

———. *Touching Liberty: Abolition, Feminism, and the Politics of the Body*. Berkeley: University of California Press, 1993.

Sanderson, George N. "The Anglo-German Agreement of 1890 and the Upper Nile." *English Historical Review* 78, no. 306 (1963): 49–72.

Schulten, Susan. *The Geographical Imagination in America, 1880–1950*. Chicago: University of Chicago Press, 2001.

———. *Mapping the Nation: History and Cartography in Nineteenth-Century America*. Chicago: University of Chicago Press, 2012.

Scott, Joan W. *Gender and the Politics of History*. New York: Columbia University Press, 1999.

———. "Gender: A Useful Category of Historical Analysis." *American Historical Review* 91, no. 5 (1986): 1053–75.

Share, Michael. *Where Empires Collided: Russian and Soviet Relations with Hong Kong, Taiwan, and Macao*. Hong Kong: Chinese University Press, 2007.

Sharkey, Heather J. *American Evangelicals in Egypt: Missionary Encounters in an Age of Empire*. Princeton, N.J.: Princeton University Press, 2008.

Shefrin, Jill. *Neatly Dissected: For the Instruction of Young Ladies and Gentlemen in the Knowledge of Geography: John Spilsbury and Early Dissected Puzzles*. Los Angeles: Cotsen Occasional Press, 1999.

———. *Such Constant Affectionate Care: Lady Charlotte Finch—Royal Governess & the Children of George III*. Los Angeles: Cotsen Occasional Press, 2003.

Sheldon, Rebekah. *The Child to Come: Life after the Human Catastrophe*. Minneapolis: University of Minnesota Press, 2016.

Shepherd, Gene D., and William B. Ragan. *Modern Elementary Curriculum*. New York: Holt, Rinehart, Winston, 1971.

Shrock, Joel. *The Gilded Age*. Westport: Greenwood Press, 2004.

Sinha, Mrinalini. *Colonial Masculinity: The "Manly Englishman" and the "Effeminate Bengali" in the Late Nineteenth Century*. Manchester: Manchester University Press, 1995.

Slocum, Jerry. "Sam Loyd's Most Successful Hoax." In *Homage to a Pied Puzzler*, edited by Ed Pegg Jr., Alan H. Schoen, Tom Rodgers, 3–22. Wellesley, Mass.: A.K. Peters, 2009. https://collections.libraries.indiana.edu/lilly/exhibitions_legacy/collections/overview/puzzle_docs/Sam_Loyd_Successful_Hoax.pdf.

Smith, Neil. *American Empire: Roosevelt's Geographer and the Prelude to Globalization*. Berkeley: University of California Press, 2003.

Smith, Susan Harris, and Melanie Dawson. *The American 1890s: A Cultural Reader*. Durham, N.C.: Duke University Press, 2000.

Smith, Woodruff D. "Friedrich Ratzel and the Origins of Lebensraum." *German Studies Review* 3, no. 1 (1980): 51–68.

Spearman, Mindy. "Race in Elementary Geography Textbooks: Examples from South Carolina, 1890–1927." In *Histories of Social Studies and Race: 1865–2000*, edited by Christine Woyshner and Chara Haeussler Bohan, 115–34. New York: Palgrave Macmillan, 2012.

Stahl, J. D. "Children's Literature." In *American History through Literature, 1870–1920*, edited by Tom Quirk and Gary Scharnhorst, 220–25. Detroit, Mich.: Charles Scribner's Sons/Thomson Gale, 2006.

Steet, Linda. *Veils and Daggers: A Century of National Geographic's Representation of the Arab World*. Philadelphia: Temple University Press, 2000.

Steinmetz, George. "Geopolitics." In *The Wiley-Blackwell Encyclopedia of Globalization*, edited by George Ritzer, 800–23. West Sussex: Wiley-Blackwell, 2012.

Stephenson Kathryn. "The Quarantine War: The Burning of the New York Marine Hospital in 1858." *Public Health Reports* 119, no. 1 (1974): 79–92.

Stoler, Ann Laura. *Carnal Knowledge and Imperial Power: Race and the Intimate in Colonial Rule*. Berkley: University of California Press, 2002.

———. "Tense and Tender Ties: The Politics of Comparison in North American History and (Post) Colonial Studies." *Journal of American History* 88, no. 3 (2001): 829–65.

Stratton, Clif. *Education for Empire: American Schools, Race, and the Paths of Good Citizenship*. Oakland: University of California Press, 2016.

Sutton-Smith, Brian. "Does Play Prepare for the Future?" In *Toys, Play, and Child Development*, edited by Jeffrey H. Goldstein, 130–46. Cambridge: Cambridge University Press, 1994.
Taylor Saito, Natsu. "Model Minority, Yellow Peril: Functions of Foreignness in the Construction of Asian American Legal Identity." *Asian American Law Journal* 4 (1997): 71–95.
Tebbel, John, and Mary Ellen Zuckerman. *The Magazine in America, 1741–1990*. New York: Oxford University Press, 1991.
Thompson, Ewa M. *Imperial Knowledge: Russian Literature and Colonialism*. Westport, Conn.: Greenwood Press, 2000.
Thongchai Winichakul. *Siam Mapped: A History of the Geo-Body of a Nation*. Honolulu: University of Hawaii Press, 1994.
Toal, Gerard. *Critical Geopolitics: The Politics of Writing Global Space*. Minneapolis: University of Minnesota Press, 1996.
Trafzer, Clifford E., Jean A. Keller, and Lorene Sisquoc, eds. *Boarding School Blues: Revisiting American Indian Educational Experiences*. Lincoln: University of Nebraska Press, 2010.
Tröhler, Daniel. *Pestalozzi and the Educationalization of the World*. New York: Palgrave Macmillan, 2013.
Tuttle, William M., Jr. *"Daddy's Gone to War": The Second World War in the Lives of America's Children*. Oxford: Oxford University Press, 1993.
Tyrrell, Ian. *Reforming the World: The Creation of America's Moral Empire*. Princeton, N.J.: Princeton University Press, 2010.
Ulanowicz, Anastasia. "Philippe Ariès." Representing Childhood, University of Pittsburgh. Accessed October 2020. www.representingchildhood.pitt.edu/aries.htm.
Vallgårda, Karen, Kristine Alexander, and Stephanie Olsen. "Against Agency." Society for the History of Children and Youth. October 23, 2018. www.shcy.org/features/commentaries/against-agency.
Verdier, Nicolas, and Jean-Marc Besse. "Color and Cartography." In *The History of Cartography*. Vol. 4, *Cartography in the European Enlightenment*, edited by Matthew H. Edney and Mary Sponberg Pedley, 294–302. Chicago: University of Chicago Press, 2020.
Vining, James W. "Astronomical Geography: An Examination of the Early American Literature." *Geographical Bulletin* 23 (May 1983): 30–40.
Wallace-Sanders, Kimberly. *Mammy: A Century of Race, Gender, and Southern Memory*. Ann Arbor: University of Michigan Press, 2008.
Watts, Sarah. *Rough Rider in the White House: Theodore Roosevelt and the Politics of Desire*. Chicago: University of Chicago Press, 2003.
Weick, Karl E. "Cartographic Myths in Organizations." In *Mapping Strategic Thought*, edited by Anne S. Huff, 1–10. Chichester: Wiley, 1990.
Weikle-Mills, Courtney. *Imaginary Citizens: Child Readers and the Limits of American Independence, 1640–1868*. Baltimore: Johns Hopkins University Press, 2013.
Weinstein, Amy. *Once Upon a Time: Illustrations from Fairytales, Fables, Primers, Pop-Ups, and Other Children's Books*. New York: Princeton Architectural Press, 2005.

Wendt, Simon, and Pablo Dominguez Andersen. *Masculinities and the Nation in the Modern World: Between Hegemony and Marginalization.* New York: Palgrave Macmillan, 2015.

Wiegand, Patrick. *Places in the Primary School: Knowledge and Understanding of Places at Key Stages 1 and 2.* London: Falmer Press, 1992.

Wilkens, Matthew. "Canons, Close Reading, and the Evolution of Method." In *Debates in the Digital Humanities,* edited by Matthew K. Gold. Minneapolis: University of Minnesota Press, 2012. https://dhdebates.gc.cuny.edu/read/untitled-88c11800-9446-469b-a3be-3fdb36bfbd1e/section/6c7cbaa1-5ff8-4439-9ffb-aeccbc6d5734.

Williams, Anne D. *Jigsaw Puzzles: An Illustrated History and Price Guide.* Radnor: Wallace-Homestead, 1990.

Withers, Charles W. J. *Placing the Enlightenment: Thinking Geographically about the Age of Reason.* Chicago: University of Chicago Press, 2007.

Witt, John Fabian. *Patriots and Cosmopolitans: Hidden Histories of American Law.* Cambridge, Mass.: Harvard University Press, 2007.

Wolpoff, Milford, and Rachel Caspari. *Race and Human Evolution: A Fatal Attraction.* New York: Simon & Schuster, 2007.

Wong, Edlie L. "Around the World and across the Board: Nellie Bly and the Geography of Games." In *American Literary Geographies: Spatial Practice and Cultural Production, 1500–1900,* edited by Martin Brückner and Hsuan L. Hsu, 296–324. Newark: University of Delaware Press, 2007.

Wood, Denis, and John Fels. *The Power of Maps.* New York: Gilford Press, 1992.

Wood, Gordon S. *Empire of Liberty: A History of the Early Republic, 1789–1815.* New York: Oxford University Press, 2009.

Woodard, Colin. *American Nations: A History of the Eleven Rival Regional Cultures of North America.* New York: Viking, 2011.

Wright, John Kirtland. *Geography in the Making: The American Geographical Society, 1851–1951.* New York: American Geographical Society, 1952.

Zahra, Tara. "Imagined Noncommunities: National Indifference as a Category of Analysis." *Slavic Review* 69, no. 1 (2010): 93–119.

Zolberg, Aristide R. *A Nation by Design: Immigration Policy in the Fashioning of America.* Cambridge, Mass.: Harvard University Press, 2006.

Tertiary Sources

Ágoston, Gábor, and Bruce Alan Masters. *Encyclopedia of the Ottoman Empire.* New York: Fact on Files, 2009.

Arnold, Kathleen R. *Anti-Immigration in the United States: A Historical Encyclopedia.* Santa Barbara: Greenwood Press, 2011.

Beede, Benjamin R., ed. *The War of 1898 and U.S. Interventions, 1898–1934: An Encyclopedia.* New York: Routledge, 1994.

Fass, Paula S. *Encyclopedia of Children and Childhood: In History and Society.* 3. vols. New York: Macmillan Reference USA, 2004.

Finkelman, Paul. *Encyclopedia of the United States in the Nineteenth Century.* New York: Charles Scribner's Sons, 2001.

Gregory, Derek, Ron Johnston, Geraldine Pratt, Michael Watts, and Sarah Whatmore. *The Dictionary of Human Geography*. Chichester: Wiley Blackwell, 2009.

Ménage, Gilles. *Les Origines de la Langue Françoise de Ménage*. Paris: Chez Augustin Courbé, 1650.

Miller, Randall M. "Industrial Age, Recreational Life." In *The Greenwood Encyclopedia of Daily Life in America*, 587–611. Westport, Conn.: Greenwood Press, 2009.

Oxford Dictionaries, Dictionary. Accessed September 2020. http://oxforddictionaries.com/definition/entanglement.

Ritzer, George. *The Wiley-Blackwell Encyclopedia of Globalization*. Chichester, West Sussex: Wiley-Blackwell, 2012.

Index

Note: Page numbers in italics refer to figures.

Abbott, Jacob, 56, 186n91; *Rollo's Travels*, 56, 145, 174n31, 186n91
ABC books, 10, 62, 110, 111–12, 116, 190n43, 194n39, 195n45. *See also* letters of the alphabet
ableism, 177n58, 201n30, 204n35
abolitionism, 183n28. *See also* African American
Adams, Abigail, 25, 29
Adams, Daniel, 30
Adirondack Mountains, 117
Adventures of Huckleberry Finn, The (Twain), 157–59, 161–62, 166, 203n28
Adventures of Tom Sawyer, The (Twain), 1–4, 161, 171n1, 171n2, 171n3, 172n10, 172n12
Africa, 28, 32, 34, 37, 40, 49, 60, 64, 73, 78–79, 92, 95–97, 101, 105–6, 115, 116, 117, 118–19, 126, 134, 150, 193n34, 195n57
African American, 9, 12, 35; and African identity, 193n34; and plantation life, 196n62; and racial violence, 196n62; and racism, 194n44, 199n12; and resistance, 107; and sexual violence, 196n62; and slavery, 2, 8, 55, 65, 106, 107, 136, 183n28, 186n88, 193n34, 196n62; in U.S. military, 83–84
age of empire, the, 3, 13, 24, 58, 108, 163, 166, 177n50
age of exploration, the, 179n71, 183n30, 186n88, 190n46, 204n33
Alabama, 99
Algeria, 113
Alphonse Isle, 99, 101–3, 109, 192n23
Al-Rashid, Haroun, 159–60, *160*, 161
Amazon, 117

American exceptionalism, 26, 34, 121, 143; history of, 181n7. *See also* patriotism
American Geographical Society, 26, 181n5
American imperialism, 12, 20, 58, 147, 159, 163, 167, 172n10, 189n36, 201n24; imperial governance, 168, 203n7; imperial pedagogy, 4, 10; institutionalization of imperial cultures, 203n7
American Institute of Instruction, 87, 191n55, 191n56
Americanization, 172n14; of native Americans, 137, 147, 200n19; resistance against, 107, 137, 199n13; un-American lifestyles, 199n13
American Revolution, 21, 25–26, 28–30, 34, 76, 181n3
American Studies, 26–27; Archipelagic American Studies, 177n50, 177n51, 177n52; InterAmerican studies, 182n9; transnational American studies, 26
Americas, the, 13–14, 16, 35, 40, 64, 76, 115, 117, 119, 126, 136, 145, 160–62, 182n9, 184n42
Amoy (China), 90, 116–17
Anderson, Benedict, 3, 58–59, 73–74, 161, 171n9; imagined communities, 161, 187n99, 187n100, 189n34, 189n35. *See also* imagined noncommunities
Anglo-German Colonial Agreement, 119
Ann Arbor (Michigan), 44
Antwerp, 149–50
Ariès, Philippe, 173n21; *Centuries of Childhood*, 173n21; critique of, 173n21
Ashfield (Sydney), 154
Asia, 13, 28, 37, 41, 53, 60, 101, 109, 115, 117, 118, 119, 126, 134, 142

Asian American, 9, 204n35
Athens, 195n59
Atkinson, Fred W., 164, 203n10, 203n11, 203n12, 203n13, 203n14, 203n15, 203n16, 203n17
Atlantic, the, 13, 29, 34, 37, 38, 40, 56, 61, 63, 74, 78, 94, 98, 173n18, 179n71, 189n36, 189n41
atlases, 16, 30, 65, 69, 71, 73, 78, 101, 108, 109, 115, 119, 124, 125, 179n71, 182n24, 183n26, 184n59
Australia, 37, 40, 73, 102, 113, 115, 117, 119, 154, 175n37, 194n33
Austria, 32, 117, 151

babes in the wood, 102, 193n24. *See also* Norfolk Tragedy
Bachelard, Gaston, 18
Bagdad, 159, 160
Baghdad, 159–60
Baltimore, 123, *123*
Battle of Manila, 82, 84, 153–54. *See also* Mock Battle of Manila
Battle of San Juan Heights, 83–84, 190n51
Baudrillard, Jean, 3, 171n8
Baum, L. Frank, 192n11; and U.S. elections, 192n11; works: *The Wonderful Wizard of Oz*, 192n11
Beecher, Catherine, 185n72
Beecher Stowe, Harriet, 30, 183n28; works: *First Geography for Children*, 30, 183n28; *Uncle Tom's Cabin*, 122, 196n62
Belgium, 150
Bennett, Jack, 160–63
Berlin, 60
Bill of Rights, 25
biopolitics, 107, 134, 142. *See also* necropolitics
Boac (Island of Marinduque), 163
boarding schools for Native American, 137, 145, 200n19; Americanization at, 137, 145, 200n19; Carlisle Indian Industrial School, 200n19; racial violence and, 137, 145, 200n19
body politics, 107, 142–44, 192n18. *See also* biopolitics

Bombay, 60
Borges, Jorge Luis, 171n6
Boston, 39, 135, 180n93, 191n56, 197n68
Bourne, Henry Eldridge, 183n30, 184n46, 184n47
Boxer Movement, 160
Boy Scouts of America, 98, 176n44; imperialism and, 176n44; masculinity and, 98, 176n44
Brazil, 150
Brigham, Albert Perry, 49–50, 172n11, 185n77, 186n78
British Empire, 10, 64–65, 76, 123, 175n37, 181n2, 196n65
British India, 56
British North America, 111
Brooklyn Daily Eagle, 94
Bryan, William Jennings, 94, 192n11
Bulletin of the American Geographical Society, 50
Burma, 117
Burnett, Frances H., 120–22, 195n60, 196n60, 196n62, 196n62; works: *Little Lord Fauntleroy*, 122, 195n60, 196n62; *The One I Knew Best of All: A Memory of the Mind of a Child*, 196n62; *The Secret Garden*, 196n59

Cambridge (England), 195n59
Canada, 145
Cape Town, 60
Caribbean, 13, 26, 56, 87, 141, 152, 163, 193n24
Carson, Rachel, 199n8
Cartography, 2–5, 14–18, 20, 23–24, 34, 37, 41, 51–52, 57–60, 62, 66–67, 69–71, 73–74, 81, 85, 88, 90, 93, 101, 107–11, 116, 121, 126, 155, 166, 169, 171n5, 175n35; empire and, 98–99, 160, 167, 178n64, 187n99, 189n36; gender and, 26, 59, 75, 80–81, 187n102, 190n48; race and, 32, 38, 39, 54, *54*, 78–79, 107, 185n61; war and, 14, 83–84, *83*, 160, 175n35, 178n59
cartography-in-progress, 22, 168–69
Central America, 101
Central School Journal, 44

Centuries of Childhood (Ariès), 173n21
Charles Scribner's Sons, 186n84
Chicago, 15
childhood: adulthood and, 10, 12, 30, 56, 68, 75, 86, 93, 102–3, 109, 128, 156–57, 168–69, 173n21, 174n23, 175n39, 175n40, 198n8; archives of, 11–12, 23, 92, 155, 167–68, 172n12, 190n43, 204n35; as artifact, 8, 10–11, 12, 62, 89, 156, 158, 174n23, 174n23, 175n36; as biosocial developmental stage, 7, 10, 174n23; creativity and, 3, 22, 73, 74, 88, 89, 124–25, 155, 158; disability and, 66; ecocriticism and, 51, 193n27; empire and, 5, 7, 10, 11, 22, 24, 50, 67, 71, 74, 92–93, 98, 105, 149, 152–53, 156, 158, 160–62, 163, 165–66, 168–69, 175n31, 175n37, 176n43, 183n24, 204n35; innocence and, 89, 175n36, 196n62; invisibility of, 175n39; as lived experience, 8, 11, 174n23, 176n43; nationhood and, 9–12, 29–30, 34, 57, 90, 122, 125, 127, 131, 158, 167, 203n5; nostalgia for, 62, 190n43; as sociocultural construct, 8–9, 89, 174n23; war and, 71, 82–84, 132, 144, 149, 152–54, 202n38
child play, 5, 8, 10, 12, 21–24, 56–59, 61–64, 66–73, 74–75, 80–81, 85, 88–89, 93, 95, 99, 104, 112, 125–28, 135, 158, 166, 174n23, 175n37, 187n7, 189n28, 189n29, 190n43, 191n62, 193n24, 196n62, 196n62, 197n68; doll-play, 62, 108, 122, 188n10, 190n43, 191n62, 196n62, 196n62; domestic play, 21, 62, 66–68, 74–75, 80, 193n24, 196n62; parlor games, 62; party games, 10; team sports, 10
children: African American children, 122, 176n43, 204n35; children's book market, 25, 186n91; children's history, 11; children's letters, 24, 90–91, 129–32, 147, 154–55, 193n27, 197n2, 198n4, 198n8; children's literature, 174n30, 193n27; children's magazines, 22–24, 90–93, 125–28, 129–32, 150, 154–58, 180n90, 180n93, 180n94, 191n1, 198n6, 199n9; child psychology, 9, 195n50; children's voice, 11, 93, 131, 133, 174n24, 198n8; colonialism and, 11, 174n30, 174n31; as consumers, 62–63; as historically present, 11–12, 175n39, 175n40; as individuals, 11; multilingualism among, 134–35, 147–48, 150, 200n15, 201n30; Native American children, 200n19, 200n20; and playfulness, 4, 74; as proto-citizens, 52, 90, 176n44; socializing of, 67, 111, 137, 167, 195n45
children's toys and games, 4, 10, 18, 20–22, 61–63, 65–68, 70–74, 75, 79–80, 87, 89, 93, 126, 131, 145, 157, 167–68, 175n37, 187n7, 188n11, 188n20, 189n28, 190n42; alphabet cards, 62; black doll, 122, 196n62, 196n62; board games, 21, 61–62, 68, 157–58, 175n37, 187n7; clay marbles, 190n43; corn husk dolls, 190n43; map games, 21–22, 62, 187n7; paper dolls, 62; pick-up sticks, 190n43; topsy-turvy, 122, 196n62
Chile, 150
China, 28, 32, 33, 49, 76, 79, 113, 115, 117–18, 120, 189n39, 196n67; Chinese Exclusion Act, 94, 192n14, 197n66; Chinese immigration to the U.S., 54–55, 94–95, 111; Chines-U.S. relations, 79, 94, 124, 189n39, 196n67
cholera scare (September 1893), 140–41
Christian missionaries, 14, 15–16, 49, 106–7, 123, 136–38, 146, 149–50, 179n67, 186n94, 200n18; childhood and, 106, 136–39, 146, 149, 186n94; empire and, 49, 106, 123, 150; race and, 107, 136–38, 146
Churchill Semple, Ellen, 59, 172n16, 173n18
Clute, John J., 30
Cochran, Elizabeth Jane (Nellie Bly), 61, 187n1
Cole, Thomas, 178n66
colleges and universities, 6–7, 44–46, 164, 184n55; Columbia University, 46, 184n55; Cornell University, 49; Harvard University, 44, 47, 50, 167, 173n18, 184n55; Stanford University, 184n55; University of Michigan, 45, 184n55; University of Pennsylvania, 184n55

colonialism, 16, 107, 171n9; European colonialism, 106
colonization, 3–4, 8–9, 11, 15–16, 22, 27, 49–51, 57, 60, 69, 74, 76, 89, 102, 104, 106–9, 118–19, 122, 152, 163–64, 171n9, 174n31, 179n77, 193n26; colonial settlement, 16, 44, 49, 58, 107, 122, 136–37, 142, 176n47, 195n56, 200n20
Committee of Ten, 7, 47, 173n20
Committee on Elementary Education, 45
Committee on Secondary School Studies, 44
Congo, the, 106–7, 185n60, 193n30, 193n30, 193n31
Congo River, 106, 193n32
Connecticut Yankee in King Arthur's Court, A (Twain), 160
Constantinople, 123, 196n65
consumer culture, 8, 12, 15, 61–63, 80, 89, 92, 106, 190n43
Cornell, Sarah S., 40, 59
cosmopolitanism, 181n4
Costa Rica, 87
Cuba, 14, 41, 60, 71, 82–86, 164, 178n60
cultural geography, 168

Davis, William Morris, 6, 46–51, 104, 172n16, 173n18, 182n15, 185n65
Dawes Act (1879), 137, 201n22
de Beaumont, Jeanne-Marie LePrince, 63–65, 189n36
De Certeau, Michel, 18
Dehra Dun (India), 149
de la Blache, Vidal, 6, 50, 186n79
Deleuze, Gilles, 88, 191n57, 191n58
Dewey, Admiral George, 153–54
digital humanities, 195n54
Dodge, Mary Mapes, 131–32, 171n1, 191n1, 198n3, 198n6; editorial views, 131–32, 191n1, 198n3, 198n6; works: *Hans Brinker, or the Silver Skates*, 198n6
Dodge, Richard Elwood, 6, 46–51, 60, 104, 172n16, 180n94, 184n59
Downes v. Bidwell, 140
Dresden, 151, 195n59

Du Bois, W. E. B., 60, 102, 187n105
Dwight, Nathaniel, 28–29, 182n16, 182n17, 182n18

Ecocriticism, 51, 193n27
Egypt, 113, 122, 171n2, 186n94
England, 8, 25, 29, 94, 113, 172n10
English Channel, 121, 122
Eurocentrism, 6–7, 34, 41, 113, 119, 126
Europe, 6, 25, 28, 34–38, 40, 50, 53, 60, 62, 64, 66, 92, 101, 104, 113–15, 117, 119, 123, 126, 146, 148, 151, 183n30, 185n60, 186n91, 188n20, 189n41; European empires, 3–4, 13, 14, 20–22, 30, 41, 56, 60, 63, 73, 76–79, 81, 85–86, 94, 101, 103, 106–7, 116, 118–19, 123–24, 177n50, 178n64, 179n71, 190n48, 192n23; European migrants to the United States, 5, 9, 52, 61, 121, 122, 134–36, 139–42, 143–44, 150, 155, 199n13, 200n14, 200n15, 201n27, 201n29

family of nations, 9, 14, 78, 127, 159, 174n29
Far East, 15, 32, 76, 97, 149, 185n60
Fewkes, Jesse Walter, 173n18
Finland, 122
First Barbary War, 32
First Geography for Children (Beecher Stowe), 30, 183n28
First Opium War, 123
Florence, 99, 108
Fort Wadsworth, 139
Foucault, Michel, 18, 126, 134
France, 7, 55, 65, 86–87, 94, 150
Frankfort (Kentucky), 195n56
Frankfurt (Germany), 118, 195n56
French Empire, 73–74, 76, 101, 109

Gender, 66, 79–80, 91, 111, 133, 167, 169, 175n36, 176n45, 178n64, 190n44, 190n47, 192n18, 196n62, 197n2; boys, 2–3, 46, 80, 98, 102, 106, 137–39, 147–49, 153–54, 162, 171n1, 171n2, 176n44, 185n60, 186n91, 197n1; fathers, 56, 75, 131, 136, 146, 181n95, 189n37; gender order, 80, 190n44; girls, 63, 90, 132, 133, 138, 191n62, 196n62,

198n4; masculinity, 26, 59, 75, 78, 80–81, 97–99, 176n44, 176n45, 187n102, 190n48, 192n18; mothers, 11, 75, 103, 131, 137–38, 181n95, 198n4; war and, 202n38. *See also* manifest domesticity

Geneva, 113, 195n50

geographic pedagogies, 4, 28, 29, 46–47, 51, 56, 70, 164–65, 183n24, 185n61. *See also* home geography

geopolitics, 4, 6, 67, 73, 166, 172n16, 175n35, 184n55, 189n36; European roots, 6, 172n16; geopolitical imaginaries, 73, 79, 87, 161, 202n4

Geopolitik. See geopolitics

George III, King, 63–65

Get off the Earth, A Puzzle Mystery (Loyd), 94, 95, 111

Ginzburg, Carlo, 20; evidential paradigm (*Paradigma indiziario*), 20. *See also* semiotics of the overlooked

Globale Mikrogeschichte. See relational global microhistory

Globes, 183n24; rotating globe, 94–95, 111; slated globes, 183n24; table globes, 101, 119, 127, 165; terrestrial globes, 37, 40, 41, 69, 103, 106–8, 182n24

Goodrich, Samuel G. (Peter Parley), 31–32, *31, 33*, 36, *36–37, 38*, 39–40, 183n37

Grand Junction (Colorado), 138

Greece, 113, 151

Guattari, Felix, 88, 191n57, 191n58

Guyot, Arnold Henry, 52–55, *54*

Haiti, 56

Hall, Granville Stanley, 8, 56, 174n27, 180n89, 186n90

Hamburg, 118

Hanover (Germany), 195n59

Hans Brinker, or the Silver Skates (Dodge), 198n6

Hanzi characters, 111, 194n43. *See also* letters of the alphabet

Harper's New Monthly Magazine, 171n1

Harper's School Geography, 52–53, *53*, 75, 186n83

Harper's Young People: An Illustrated Weekly: Harper's Round Table, 22–23, 90–93, 99, *100*, 102–5, *103, 105*, 110, 112–16, *114, 115, 117*, 119–21, *120*, 123, *124*, 124–26, 129–35, 150, 180n94, 191n1

Harvey, David, 17–20, *19*, 69, 108, *109*, 179n70, 179n78, 179n79, 180n80, 180n81, 189n30, 194n35

Hawai'i islands, 41, 146–48, 172n10; agricultural products, 147; United States and, 146–48, 164, 166, 202n35

Helgoland-Sansibar-Vertrag. *See* Anglo-German Colonial Agreement

Henderson, Charles Richmond, 173n18

Hill, Samuel L., 189n40

historical childhood studies, 11, 12, 166, 173n21, 174n24

history of emotions, 11

Hoffman Island, 140

Holy Land, 15

Home: aspirational visions of, 133–34, 154, 197n73; home front, 51, 149–51, 155, 202n38; the nation as, 14, 15, 24, 28, 47–50, 51, 55, 59, 70, 89, 92, 102–4, 110, 112–13, 121, 128, 129, 131, 133–34, 135–36, 138, 141, 143–45, 147, 150–52, 155, 162, 169, 186n80, 194n37, 195n50, 197n73; non-home, 89, 130–31, 133–34, 136, 145, 147, 155; as safe haven, 49, 68, 103, 150, 193n24. *See also* home geography

home geography, 21, 23, 28, 46–52, 55–56, 59, 66, 70, 102, 104–5, 122, 129, 133, 145, 147, 152, 157, 165, 182n14, 184n59, 185n61, 186n82

homonymy, 89, 92, 99, 101–4, 108, 122, 193n25; homographs, 193n25; homophones, 193n25

Hong Kong, 123, 196n65

Hubbard, George D., 49–50, 185n75, 185n76

Hudson (New York), 143

Hudson Bay, 122

human geography, 49, 173n18, 179n75, 204n31

Iconic Turn, the, 191n52

Idaho, 99

Illinois, 1, 99, 135
Imagined Noncommunities, 203n29, 204n34. *See also* imagined Noncommunities
Immigration, 13, 53, 142–44, 200n14; anti-immigration policies, 94, 141, 196n67, 200n14; anti-immigration sentiments, 53, 111, 133–34, 143–44, 200n14, 202n33; Chinese immigrants, 94–95, 95, 111, 192n14, 192n18, 196n67; Filipino immigrants, 192n18; Irish immigrants, 201n29; nativism and, 13, 26, 53, 111, 143–44, 172n10, 200n14, 201n30, 202n33; Swedish immigrants, 200n13. *See also* China: Chinese Exclusion Act; National Quarantine Act
Imperial Germany, 11, 76
Imperial Turn, the, 203n7
Indiana, 1–3
Indian Ocean, the, 99, 101
Innocence Abroad (Game), 61
Innocents Abroad, or the New Pilgrim's Progress (Twain), 61, 178n65, 187n6
intersectionality, 32, 169, 190n47
Ireland, 101, 118, 185n60
Isle of Man, 101
Ispahan (Isfahan), 28
Italy, 87, 113
Ives, S. B., 61, 189n40

Jamaica, 64
Jameson, Frederic, 15–17, 108, 179n68, 179n70, 179n72, 179n74, 179n76, 202n42
Japan, 32, 54, 79, 86–87, 95, 194n43
Jefferys, Thomas, 63–65
Jernegan, Prescott Ford, 165–66; and imperial pedagogy, 165–66
jigsaw effect, 74, 189n36
Journal of School Geography, 46–47

Kentucky, 195n56
Kiel Canal, 119
Kilimanjaro, 118
King, Charles F., 58
"King Leopold's Soliloquy" (Twain), 172n10
Kirchwey, Clara Barbara, 48–50, 60

Kjellen, Rudolf, 172n16
Knox, Thomas Wallace, 46, 106, 185n60; works: *The Boy Travellers in Australia*, 185n60; *The Boy Travellers in the Far East*, 185n60; *The Boy Travellers in Great Britain and Ireland*, 185n60; *The Boy Travellers in Mexico*, 185n60; *The Boy Travellers in Central Europe*, 185n60; *The Boy Travellers in Northern Europe*, 185n60; *The Boy Travellers in Southern Europe*, 185n60; *The Boy Travellers in the Russian Empire*, 185n60; *The Boy Travellers in South America*, 185n60; *The Boy Travellers in the Levant*, 185n60; *The Boy Travellers on the Congo*, 106
Kohala (Hawaii), 147

Landschaft geography. *See* regional geography
late capitalism, 17
Lausanne, 150
Lefebvre, Henri, 18
Lehnerts, Edward M., 184n59
Leopold II, King, 106
Letter-Box, the, 130, 132, 144, 152
letters of the alphabet: ABC books, 10, 62, 111–12, 116, 190n43, 195n45; geography and, 91–92, 99, 110–11, 112, 113, 116–17, 122–23, 126; Hanzi characters, 194n43; literacy and, 10, 110, 112, 123–24, 126, 135, 194n42, 195n54; spelling and, 10, 106, 110, 118, 120–22, 123–25, 139, 168, 193n25, 195n56; the world and, 92, 99, 112, 116–17, 122, 124–25, 127
Library of Congress, 14; Hall of Maps and Charts, 14, 178n62
literacy: child education, 10, 110, 125, 126–27, 155, 182n22, 195n54; geographic literacy, 3–5, 15–16, 20–24, 27, 30–32, 36, 41, 46–49, 51–52, 53, 55, 60, 66–67, 69–71, 73, 86–88, 90, 92–93, 100–102, 104, 108–10, 112, 115–17, 125, 178n59, 182n24; history of, 127, 194n42; imperial literacy, 4; as privilege, 24, 125, 127, 135, 139, 146–47, 155, 182n22, 194n42, 200n15

Little Lord Fauntleroy (Burnett), 122, 195n60, 196n62
Liverpool, 195n57
Livingstone, David, 106
Lockport (Illinois), 135
Lodge, Henry Cabot, 177n53
London, 28, 60, 63, 65, 121, 122, 196n59, 200n20
Long Island, 196n59
Lost 'JAP,' The, 94–97, 96
Loyd, Samuel, 94, 95, 96–98, 96, 111, 192n10; disappearance puzzles, 94, 95, 96, 97, 98–99, 111, 202n36; politics and, 94, 111, 192n11, 194n44; puzzle hoaxes, 197n68; puzzle making, 94–98, 111, 192n20, 197n68
Lynch, Kevin, 17, 179n74, 179n75

Mahan, Alfred T., 172n16, 173n18
Malte-Brun, Conrad, 78, 189n38
manifest destiny, 13, 181n7, 185n72, 190n48. *See also* manifest domesticity
manifest domesticity, 80, 99, 103, 181n7, 185n72, 190n46, 190n48
Manila, 82, 84, 153, 154, 163, 165–66, 202n41, 203n18. *See also* Battle of Manila; Mock Battle of Manila
Mann, Horace, 185n72
mapping. *See* cartography
Maps: cartographic maps, 2–3, 16–18, 20, 66, 69, 71; cognitive maps, 16–18, 20, 22, 24, 27, 57, 67, 69–70, 108, 116–17, 125–26, 128, 155–56, 158, 168–69, 179n75, 179n77; dissected maps, 10, 21–22, 62–71, 68, 73–76, 81–82, 82, 83–84, 87–88, 119, 126, 157, 188n16, 189n40; dot maps, 125; map games, 21–22, 62, 187n7; mental maps, 2, 16–17, 85, 130, 135, 149, 179n74, 179n75; outline map, 30, 183n26; puzzle maps, 63; sketch map, 57–58, 93, 155, 187n97; world map, 2, 5, 27, 34, 40–41, 63, 71, 73, 81, 92, 99, 103, 106, 108–9, 116–17, 120, 125, 126, 129, 157–58, 163, 165, 203n6
Marine Hospital Service, 140
McKinley, William, 13, 14, 87, 94, 160, 174n29, 177n53, 177n54, 177n55, 177n56, 177n57, 192n13

McLoughlin Brothers, 61–62, 65–66, 81–84, 82, 87, 187n7, 187n9, 190n43
Mercator projection, 16, 37, 40, 41, 117, 179n71
metanarrativization, 66–67
Mexico, 29, 145, 148–49, 152, 185n60, 202n40
Mexico City, 148
Microhistory, 20, 157, 180n86. *See also* Ginzburg, Carlo: evidential paradigm (*Paradigma indiziario*); relational global microhistory
Milton Bradley, 62, 65, 67, 79, 190n42, 190n43
Mitchell, Samuel Augustus, 78, 189n38
Mock Battle of Manila, 154. *See also* Battle of Manila
Morgan, Hank, 160
Morocco, 86
Morse, Jedidiah, 25–26, 28–29, 181n2, 181n3, 183n37
motherhood, 11, 75, 103, 131, 137–38, 198n4

National Education Association, 44
National Geographic Magazine, 14, 178n60
National Quarantine Act, 141, 201n26
Native Americans, 26, 29, 80–81, 105, 107, 134, 136–39, 143, 176n43, 190n43; Apache, 137, 139, 193n34; and forced removal, 26, 29, 80–81, 136–38, 195n56; and land allotment, 136–38, 201n22; Mescalero Apache, 137; the Oneida, 136, 200n18; La Paz tribe, 138; and reservation life, 133–34, 136–39, 145, 200n18, 200n19. *See also* biopolitics; boarding schools for Native American; Dawes Act (1879)
Necropolitics, 94–95. *See also* biopolitics
New England, 25, 65, 105, 106, 160
New England Primer, 111
Newfoundland, 101, 175n33
New York, 22, 60, 65, 71, 104, 119, 132, 136, 139, 140–41, 143, 174n26, 180n93, 182n16, 201n29
New York Age, 84

New York City Department of Health, 139–40
New York World, 61
New Zealand, 133
Nile, the, 106
Niles, Sanford, 39, 184n40
Norfolk Tragedy, 193n24. *See also* babes in the wood
North Africa's Barbary, 32
North America, 5, 15, 28, 37, 40, 44, 65, 71, 76, 105, 111, 121, 129, 141, 163, 164, 166, 175n37, 193n34
North Pole, 37, 62
North Sea, the, 119
Norway, 87, 113
Nova Scotia, 28

Oceania, the, 126
Ohio, 41, 99, 115, 116, 118, 120, 195n59
One I Knew Best of All: A Memory of the Mind of a Child, The (Burnett), 196n62
open-door policy, the, 57, 94, 124, 189n39, 196n67
O'Sullivan, John L., 29, 181n7, 182n23, 185n72
Ottoman Empire, the, 123, 196n65
Our Puzzle Magazine (Loyd), 94

Pacific, the, 13, 22, 26, 41, 44, 56, 87, 147, 163, 165, 193n24
Panic of 1893, the, 94
Paris, 60, 65, 151
Parker Brothers, 61–62, 72
Parkhurst, V. S. W., 189n40
Parochialism, 29–30, 34, 38, 46, 50, 57–58, 161, 181n4, 182n14, 197n73. *See also* westward expansion
Patriotism, 10, 21, 24–26, 28–31, 40, 87, 112, 122, 143–44, 151–52, 159, 164, 166, 169, 182n20, 182n22, 183n26, 191n53, 191n54, 200n14. *See also* American exceptionalism; Immigration: nativism
Peace, 14, 85, 86; arbitration, 86–87, 191n53, 191n55, 191n56; global dimensions of, 164, 177n53; inter-imperial rivalries, 14, 85–87,

164; peace education, 87; peace work, 191n55, 191n56
Pekin (Beijing), 28
Persia, 113, 115, 118, 149, 160–62
Philadelphia, 28, 99, 101, 123, 141
Philadelphia Epidemics of 1699, 141
Philippines, the, 14, 41, 71, 72, 82, 152, 158, 160, 163–66, 174n29, 202n41, 203n18
Piaget, Jean, 109, 194n37, 195n50; criticism of, 109, 194n37, 195n50; on children's cognition, 109, 112–13, 194n37, 195n50, 203n5
planetary consciousness (Pratt), 107, 182n13, 193n33
Poland, 102
polygenism, 55, 186n88
population management, 192n18. *See also* biopolitics
Poros (Greece), 151
Portugal, 87, 113
postcolonial studies, 9, 193n26
Pratt, Mary Louise, 107; planetary consciousness, 107, 182n13, 193n33
Prussia, 32
public letters, 132, 146, 155
public libraries, 199n9
public schooling, 87, 199n13, 200n15
Puerto Rico, 140, 164, 166
Puzzle of Teddy and the Lion (Loyd), 95–97, 96
Puzzles: child-made geographical puzzles, 20, 22–24, 90–94, 98–99, 101–4, *103*, *105*, 106, 108–9, 110–25, *114*, *115*, *117*, 120, 123, 124, *124*, 126–28, 155, 158, 167, 169, 192n21, 193n24, 196n60, 196n65; disappearance puzzles, 94–95, 98–99, 111, 202n36; jigsaw puzzles, 18, 21, 62–63, 66–67, 74, 83, 86, 188n12, 188n17, 188n18, 188n20, 189n36, 189n40; mathematical puzzles, 98, 192n20; meaning of in human life, 98; metamorphosis puzzles, 65, 188n18; National Puzzlers' League, 94; picture puzzles, 10, 21, 62–63, 65, 66, 79–82, 82, 83–85, 87, 188n17; puzzle maps, 63; *Puzzles from Young Contributors*, 91, 93

quarantine stations, 22, 133–34, 139–45, 150, 158, 201n26, 201n27, 201n28, 201n29, 201n31

race: Anglo-Saxonism, 102, 122, 135; racial categories, 8, 32, 38, 39, 54, 54–55, 79, 84, 98, 172n10; racial invisibility, 199n12; racial violence, 5, 95, 101, 103, 108, 111, 122, 137, 176n47, 196n62, 201n29; racialization, 24, 49, 78, 147, 192n18; racism, 32, 94, 97, 136, 141, 186n88, 189n41, 194n44, 199n12. *See also* African American; Asian American; racisms; slavery; whiteness
racisms, 199n12. *See also* polygenism; race; whiteness
Rand McNally, 40, 184n59
Ratzel, Friedrich, 6, 172n16, 173n18, 186n79, 187n103
regional geography, 50. *See also* home geography
reinscriptive cartography, 66–67. *See also* cartography
relational global microhistory, 20. *See also* microhistory
re-naming, 15, 88, 103–4, 107–9; colonization and, 18, 88, 101–2, 106–7, 108; missionary work, 107, 136–38; of places, 101–3, 104–7, 193n32; of people, 104, 107, 139; resistance to, 107, 137, 193n26, 193n34; violence and, 103–4, 109, 201n21
Rio Grande, 152
Ripley, William Z., 173n18
Roddy, H. Justin, 40–41, 43, 184n43
Rome, 8, 151
Roosevelt, Theodore R., 83–84, 87, 95–98, 173n18, 177n53, 190n50, 190n50
Root, Elihu, 87
Round the World in Eighty Days (Verne), 61
Rush, Benjamin, 25
Russia, 29, 32, 76, 113, 115, 123, 185n60
Russo-Japanese War, the, 86–87

Said, Edward, 9, 13, 58–59, 93, 177n50, 187n98, 187n99, 192n8
Samoa, 166
Santa Elena, 202n40

schoolbooks, 7, 21, 25, 28–30, 32, 78, 158, 163–64, 182n20, 183n27, 183n37, 189n38
school geography, 4, 14, 28, 30, 34–35, 38, 40–41, 44–47, 52–53, 88–89, 93, 114, 119, 126, 128, 131, 172n11, 175n35, 183n28, 183n37, 184n41, 186n83, 189n36; astronomical geography, 34–38, 40, 183n37, 184n38; commercial geography, 55, 165; mathematical geography, 39; physical geography, 34–35, 37, 50, 184n42, 185n77, 186n78
school geography, 4, 14, 34–35, 38, 40–41, 44–45, 47, 52–53, 75, 88–89, 93, 114, 119, 126, 128, 131, 175n35, 183n28, 183n37; modernization of, 6, 21, 26–28, 45–47, 52–53, 158, 175n35, 182n15, 183n37
Scotland, 101, 102
Scribner's Monthly, 23, 131, 180n95
Secret Garden, The (Burnett), 196n59
Selchow and Righter, 62
semiotics of the overlooked, 20, 22, 157. *See also* Ginzburg, Carlo: evidential paradigm (*Paradigma indiziario*)
Sexuality, 173n21; asexuality, 196n62; sexual violence, 196n62
Seychelles Islands, 192n23
Siberia, 32
sign language, 201n30; Bell, Alexander Graham, 201n30; campaign against, 201n30; nativism and, 201n30
slavery, 2, 8, 9, 55, 98, 106, 107, 122, 135–36, 183n28, 186n88, 193n34, 196n62; and its abolition, 183n28; and its legacy, 80, 83–84, 107, 122, 134, 138, 196n62; plantation life, 196n62. *See also* necropolitics; polygenism
Smith, J. Russell, 173n18
South America, 32, 37, 102, 116, 185n60
Southeast Asia, 152
Spain, 76, 85, 87, 164
Spanish-American War, 132, 144, 154, 177n53, 187n7. *See also* Spanish-Cuban-Philippine-American War
Spanish-Cuban-Philippine-American War, 14, 41, 60, 71, 82–84, 86, 178n60. *See also* Spanish-American War
Spanish Empire, the, 152, 165

Spilsbury, John, 63–65, *64*, 188n16, 189n36
Stanley, Henry Morton, 106, 108, 119, 193n29, 193n30, 193n30, 193n32
Stanley Falls (Boyoma Falls), 106–7; re-naming of, 106, 193n32
Star-Spangled Banner, 144
Staten Island, 139–42, 201n27
St. Nicholas: Scribner's Illustrated Magazine for Girls and Boys, 23, 129–32, 134–36, 138–39, 143–44, 146–54, 159–60, 171n1, 180n94, 180n95, 191n1, 195n60, 196n62, 198n6, 198n8, 199n9, 202n40. *See also* Dodge, Mary Mapes
St. Nicholas League, 199n8
Sutton-Smith, Brian, 68, 189n29
Sweden, 87
Swinburne Island, 140
Swinton, William, 40, 184n41
Switzerland, 113, 149–50
synecdoche, 89, 92, 99, 110, 112, 122

Tangier, 149
Teller Institute, 138
Texas, 29, 141; annexation of, 29
Texas Institute for the Blind, 66, 188n24
"The War Prayer" (Twain), 172n10
Tokyo, 122
Tom Sawyer, Abroad, Tom Sawyer Detective, and Other Stories (Twain), 171n1
Tom Sawyer, Detective (Twain), 171n1
Tom Sawyer Abroad (Twain), 1–4, 157, 161, 171n1, 171n2, 171n3, 172n10, 172n12
toponyms (geographical place names), 3, 23, 30–31, 89, 91–93, 99, 103–4, 108–10, 112–18, 120–23, 125–27, 168–69, 193n33, 195n59. *See also* re-naming
"To the Person Sitting in Darkness" (Twain), 172n10
tourist age, the, 15
toy books, 62. *See also* children's toys and games
toy manufacturers, 4–5, 65–67, 71–73, 79–80, 87, 190n43
Turkey, 76, 116, 119

Turner, Frederick Jackson, 6
Twain, Mark (Samuel Langhorne Clemens), 2–3, 15–16, 61, 158, 161, 171n1, 172n10; autobiography, 171n1, 172n10; works: *Adventures of Huckleberry Finn*, 159, 162, 203n28; *A Connecticut Yankee in King Arthur's Court*, 160; *The Adventures of Tom Sawyer*, 171n1; "King Leopold's Soliloquy," 172n10; *The Innocents Abroad, or the New Pilgrim's Progress*, 15–16, 61; "The War Prayer," 172n10; "To the Person Sitting in Darkness," 172n10; *Tom Sawyer Abroad*, 1–4, 161, 171n1, 171n2, 171n3, 172n10; *Tom Sawyer, Abroad, Tom Sawyer Detective, and Other Stories*, 171n1; *Tom Sawyer, Detective*, 171n1

Uncle Tom's Cabin (Beecher Stowe), 122, 196n62
United Provinces, 64
United States Capitol, 81, *82*
United States Census Bureau, 14, 189n28
United States Geological Survey, 26
United States Philippine Commission, 163–64, 203n8, 203n9
U.S. Congress, 5, 201n22
U.S. Constitution, 8–9, 25
U.S. Empire, 3, 4–7, 9–11, 12, 13–14, 18, 21–22, 24, 26–28, 41, 50, 58, 60, 67, 71, 73–74, 76, 78–79, 85, 89, 92–94, 98, 102, 105, 107–8, 123–24, 140, 149, 152, 156, 157–58, 160, 162–63, 165–66, 169, 172n13, 174n29, 177n50, 179n69, 181n3, 182n24, 186n94, 190n46, 192n18, 202n35; and Anti-Imperialist League, 172n10; encounters with European empires, 13–14, 41, 60, 71, 76–79, 81, 82–86, 94, 123, 124, 132, 144, 152, 154, 165, 177n50, 177n53, 178n60, 196n65; and frontier crisis, 5, 16, 56, 66, 80, 86, 98, 105, 118, 157; overseas expansion of, 13–14, 20, 21, 58, 60, 71, 78–79, 87–88, 98, 102, 104–5, 122, 149, 151, 157, 174n29; and spatial tension, 5, 14, 150; as a spatially unsettled

empire, 5–6, 13, 157, 203n6, 204n35; and territorial expansion, 9, 12, 13, 29, 80–81, 145, 157, 177n50; westward expansion, 5, 29, 41, 49, 71, 78–81, 174n29, 193n24

Verne, Jules, 61, 171n1
Vienna, 195n59
Virginia, 65, 99, 120, 184n55
Virginia Company of London, 200n20
von Humboldt, Alexander, 6

Wagner, Thomas, 189n40
War of Independence, 25
Warsaw, 118, 195n59
Washington, D.C., 81, 82
Webster, Noah, 25, 29, 181n1
Weimar, 22, 152
Western Europe, 37, 50, 101
Western Frontier, 16, 80, 145–46, 150, 157; the closing of, 5, 56, 66, 86, 98, 105, 118, 157; the frontier crisis, 172n14
West India Islands, 64
westward expansion, 5, 41, 49, 71, 78, 79, 80, 174n29, 193n24
whiteness, 2, 4, 7–9, 11, 21, 29–30, 32, 38, 49, 52–55, 57, 62, 74, 78–81, 83–84, 94, 97–98, 104, 107, 111, 122, 128, 134–39, 143, 145–48, 150, 163, 176n47, 177n49, 177n58, 190n43, 196n62, 197n73, 199n11, 199n12, 201n30, 204n35. *See also* African American; Asian American; Native Americans; race
Wild West Show, 79, 81, 189n41
Willard, Emma, 36, 46, 182n14
Winthrop, John, 25, 181n4, 181n7
Wisconsin, 136, 171n1
Wonderful Wizard of Oz, The (Baum), 192n11
Woodbridge, William Channing, 35, 36, 46, 182n14
world geography, 4–5, 21, 46–51, 63, 66, 70, 75, 78, 92, 93, 102–6, 108–12, 114, 123, 126–27, 129, 158–62, 169, 178n59, 183n24, 185n59
World's Columbian Exposition, 15
world travel, 2–3, 21, 61–62, 98, 101–2, 118, 187n3

xenophobia, 111. *See also* racisms

Yale, 25, 184n55
young America, 174n26
Youth's Companion, 23, 180n93

Zanesville, 90, 116–18, 117, 135, 195n59
Zanzibar, 106, 113, 118–19
Zimbabwe, 195n57